MODERN CAKE
DECORATION

Second Edition

Modern Cake Decoration

Second Edition

by

L. J. HANNEMAN, F.Inst.B.B.

Head of Department Service Industries,
Lancaster and Morecambe College of Further Education

APPLIED SCIENCE PUBLISHERS LTD
LONDON

APPLIED SCIENCE PUBLISHERS LTD
BARKING, ESSEX, ENGLAND

First Edition published 1964
Reprinted 1970
Second Edition published 1978

British Library Cataloguing in Publication Data

Hanneman, Leonard John
    Modern cake decoration.—2nd ed.
    1. Cake decorating.   2. Icings, Cake
    I. Title
    641.8′653   TX771

    ISBN 0-85334-785-9

WITH 232 ILLUSTRATIONS

© APPLIED SCIENCE PUBLISHERS LTD 1978

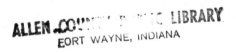
Printed in Great Britain by Galliard (Printers) Ltd, Great Yarmouth

FOREWORD TO FIRST EDITION

OVER the last few years it has been my pleasure to write many articles for the various journals of Maclaren and Sons, Ltd., on various aspects of cake decoration. It became apparent recently that this material was so comprehensive that it justified forming the basis of a book: the result is a publication which covers most aspects of cake decoration from fancies to wedding cakes. Care has been taken not to duplicate work already published, except where such work can be used as a foundation for a particular technique. A modern approach has been made to this subject and many aids to the quick commercial finishes of all types of cakes are freely used.

My grateful thanks are given to all who have helped me in the preparation of this book. I wish to record my thanks to the Principal and Governors of the Cambridgeshire College of Arts and Technology, who allowed me the use of the college facilities for preparing the many examples of work illustrated, to Mr. A. Davies, of Maclaren and Sons, Ltd., for the many excellent photographs of this work, and to John Figgins, editor, for his helpful advice.

Lastly, but by no means least, to my long-suffering wife for her encouragement, co-operation and assistance during the preparation of this book for publication.

L. J. HANNEMAN

Cambridge, 1963

FOREWORD TO SECOND EDITION

SINCE this book was first published there has been considerable interest shown in the techniques illustrated and explained and several short courses were offered by the author at which these were taught.

It was obvious that from the number of inquiries made since the first edition went out of print, a second edition would find a ready market. In producing this new edition, advantage has been taken to add new material which it is hoped will improve this already standard work.

L. J. HANNEMAN

Lancaster, 1978

CONTENTS

Chapter		Page
I.	PRINCIPLES OF DESIGN	1
II.	PREFABRICATED DECORATIONS (1)	11
III.	PREFABRICATED DECORATIONS (2)—STENCILS	22
IV.	FANCIES (1)	30
V.	FANCIES (2)—STENCILLED TOPS	46
VI.	GATEAUX (1)	62
VII.	GATEAUX (2)—STENCILLED INSCRIPTIONS AND GATEAU TOPS	68
VIII.	GATEAUX (3)—CHRISTMAS	100
IX.	TORTEN	118
X.	BATTENBURGS, LAYER CAKES, SLICES AND MARBLING	134
XI.	COMMERCIAL PIPED DECORATIONS	155
XII.	DECORATED RUN-OUTS	175
XIII.	CHRISTMAS CAKES—USE OF RUN-OUTS	186
XIV.	BIRTHDAY CAKES	203
XV.	WEDDING CAKES	210
	INDEX	231

Chapter I

Principles of Design

GOOD decoration and design demands the consideration of four factors. These are as follows:

Colour

Confectioners are restricted to what are termed 'edible colours', *i.e.* pastel tints of colours which occur naturally in foodstuffs generally. Nevertheless, other stronger colours may be introduced with advantage in order to achieve some contrast of colour in a design. The emphasis here is not so much on the choice of the contrasting colour but the strength in which it is used and the amount and manner in which it is applied.

Where very strong colour is necessary, it is useful to incorporate it into the design by means of a natural form, such as a flower, fruit, or foliage. A vivid red marzipan or sugar-paste rose, for example, is certainly the best way to introduce this colour into a design.

Green, which is always a difficult colour to use to good effect, may always be applied in the form of leaves, along with some floral arrangement. Strong, contrasting colours, used in this way, are pleasing because the eye associates such colour with the form used and not the actual cake.

In the case of celebration cakes, executed in royal icing, colour must be applied either to the actual icing or sprayed on to the cake after it has been iced. For this technique, the aerograph spray brush and equipment is necessary and will be described later in this chapter.

Gateaux, torten, fancies, etc., however, need not suffer this limitation in the application of colour, since there are so many natural decorating materials, *i.e.* cherries and angelica, that may be used which, in addition, impart a flavour and will also enhance a design with colour. Moreover, since such colours are usually very strong, they have a useful function in providing contrast. Nor should colour in this sense be restricted to the colours of the spectrum. Chocolate, both plain and milk, the almond, blanched and unblanched, or the pale cream of a genuine buttercream may also play a part in providing colour.

Consideration should also be given to light and the amount of surface reflection. A decoration which imparts a gloss will look vastly different from one which has a matt surface: yet both types may be used to advantage.

Texture

The contrast of light and shade, which may be illustrated by the difference between glossy and matt surfaces, is termed texture. The correct use of this is as important in a good design as colour. There are many techniques which may be employed to effect texture. Buttercreams may be ribbed with a serrated scraper, paddled and rippled with a knife, or piped in a variety of different forms. Almond and sugar pastes may also receive texture with different types of rolling-pins. Even the use of coralettes, sugar, etc., all helps to create highlights and shadows which add to the attractiveness of the design.

The blending of the various textures in a design is also of importance. For example, the curved line, relieved by a straight line, gives a pleasing, harmonious effect. Even a series of horizontal lines is more effective if cut by a vertical line, preferably off-centre. Contrast of line and form are of equal importance and should be implemented in every design.

Layout and Design

To a certain extent layout and design are governed by the shape and size of the cake to be decorated, but there are some guiding principles involved.

The aim should be simplicity. A design

neatly executed with the minimum of work will always look more effective than a design which is elaborate, however expertly executed: this is, of course, a personal opinion. Even so the author cannot claim to have always observed this principle himself and in the examples of gateaux shown, there are many which readers might claim to be 'elaborate'. Since, however, this is by nature a textbook, it is claimed that there is some justification in illustrating elaborate designs, even if it is only useful as a comparison for the reader.

Another important principle is to make the design attract the eye of the viewer, then by line or form bring his eye to a focal point. This may be the inscription or a centrepiece or a fruit.

Edibility

For a cake to be attractive to a customer, it must look 'delicious'. However clever the design or expertly executed, it will never look 'delicious' unless it also looks edible. Anyone who has seen the adroit way that marzipan can be shaped and coloured to imitate a piece of furniture will know what is meant. However clever it may be to see a marzipan cabinet, or a TV set, it could hardly be called 'edible', even though it may be made from the finest ground almonds and sugar. This is because such man-made objects are comparatively easy to model and the mind cannot associate a TV set, for example, with marzipan but only with wood or plastic, which in no sense can be accepted as edible. Nature is not so easily imitated and it is a mistake to try.

Here, it is only an impression that we need to create and, by so doing, the customer's mind is taken away from the natural form and accepts the material such as marzipan, sugar, or chocolate, etc., in which the impression is made. Thus, for example, the customer accepts a chocolate flower, not because of its form, but because it is chocolate and is not at all worried by the thought that such a flower is not found in Nature.

The confectioner, in fact, has unlimited scope to mutilate forms to suit his purpose, provided he does it in a way which does not detract from the edibility of the material. He can make flowers of any design, use half-flowers, and have them growing in the most unlikely places without offending the eye. But let him try to imitate Nature and the result will be a dismal failure.

Strangely enough, the mind seems to accept the imitation of colour more than it does of form. Perhaps this is because colour does not have dimension except in terms of light and shade and thus the imitation is much easier to accept.

Specialized Equipment

Many of the examples shown in this book were executed with the aid of equipment which might be termed specialized and is shown in **Figs. 1–4**.

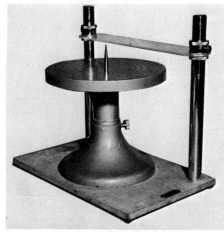

Fig. 1

Precision Turntable and Icing Aid

Fig. 1 – A considerable amount of the work illustrated has been done with a precision turntable and icing aid. The design of this equipment is unimportant, but what is essential is that the turntable should be free to revolve smoothly at the touch of a finger and to rotate without any surface wobble. Further, the turntable should accommodate a spike which should be in the absolute centre.

The icing aid is merely a rest which can be raised or lowered according to the work in hand. Again, the actual design is not important and it can be easily made from wood or metal by any handyman.

Aerograph Equipment

Fig. 2 – This consists of an air-brush which is connected by braided hose to an air supply of about 30 lb. per sq. in. pressure.

The air-brush is a pen-shaped instrument which has a small cup into which is put liquid colour. A small lever operated by the index finger controls the flow of air through the colour and also the quantity of colour allowed to emerge in the spray. By controlling this flow and the distance the pen is held from the colour is being completely atomized to form the fine spray.

After use, or between changes of colour, it is essential to wash out the air-brush thoroughly with clean water. Colour left in to dry will clog the nozzle and might result in damaging the delicate mechanism.

The air supply to the brush may be supplied from a pressure reservoir which is kept supplied either from a foot-pump or an electric compressor.

Fig. 2

object to be coloured, colour may be applied as delicately or intensely as required.

For large areas, the air-brush is held at some distance from the object to be coloured but for very small areas the brush must be held close.

It is important to use a liquid colour of the right viscosity and it is sometimes necessary to dilute colour with water (or alcohol) to bring it to the correct fluid state. One colour which invariably requires this treatment is french pink. If this is not done, the colour will not pass through the small needle valve and will not become atomized as a spray. A trial run is always advised before putting the spray into use, in order to ensure that the

Where several brushes are in use, an automatically controlled electric compressor is essential to ensure an adequate supply of air at a constant pressure.

Caramel Cutter

Fig. 3 – This consists of a shaft on which is mounted a number of circular cutting discs. These are separated by means of spacing reels of various thicknesses, so that the cutting discs may be spaced according to requirements. Handles at each end of the shaft enable the series of cutters to be rolled easily over the material to be divided. It is important to keep the cutting edge of these discs

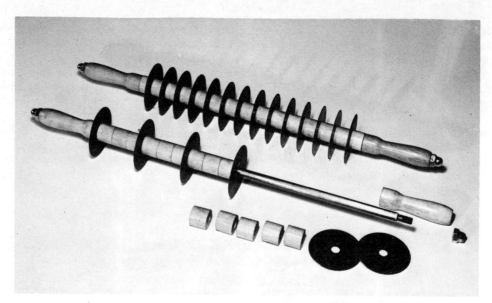

Fig. 3

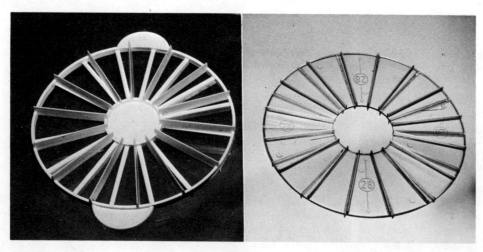

Fig. 4

4

sharp and free from a build-up of sugar or paste. Rust must also be guarded against.

Plastic Torten Markers

Fig. 4 – These may be obtained in two designs as illustrated. They are designed to mark the top of a torten into divisions, ranging from 12 to 18, according to requirements.

Preparation and Use of Decorating Materials

Fondant – This is purchased in cartons or tins of 28 lb. and is a white mass, slightly plastic. To use it as an icing, it must be reduced to a fluid consistency with stock syrup or water and heat. If the gloss is to be maintained on the icing for any length of time, it is important not to get the fondant too hot and to use a simple sugar syrup, *i.e.* 1 lb. sugar to 1 pt. water, to reduce the icing to the correct consistency. The fondant should never be in contact with fierce heat and a bain-marie, or a water jacket, should be used.

Fondant is a syrup in which there are thousands of minute sugar crystals in solution. The smaller these are the more light will be reflected and hence the better the gloss. In addition to the careful preparation of the icing, there is a number of other techniques which may be employed to preserve the gloss of fondant-covered goods. Some of these are as follows:

1 – Place the fondant-covered article into a hot oven for a few minutes. The heat of the oven will make the sugar crystals on the surface crytallize rapidly into fine crystals which will maintain a gloss for a long while.

2 – The cake base should be prepared by covering with marzipan and/or boiled puree.

3 – There may be added other materials containing jellying agents such as marshmallows, piping jelly, gelatine, or agar solution, and egg whites.

Many competitors at exhibitions resort to the use of these methods to preserve the gloss. It is important to remember, however, that the fondant should be of first quality. The gloss will never be as good if an inferior fondant is used, however much preparation is given to it.

The consistency of the icing is also of great importance. For covering large areas, the fondant should be warm and of such consistency that it can be poured over the area, spread with a palette knife, and left to leave a smooth, flat surface. For piping, however, the fondant must be of piping consistency and used almost cold.

For-high quality icings, a suitable flavour should be added to the fondant. For chocolate, the use of unsweetened chocolate or block cocoa is recommended: for coffee, a coffee extract: whilst for fruit flavours, concentrates or juices of the fresh fruit should be used.

For the best work, always use fresh fondant since, each time it is reheated, the gloss is reduced.

Royal Icing

The basic recipe for this is as follows:

7 lb. icing sugar;
1 pt. egg whites or albumen solution

The strength of the albumen solution may be varied to suit individual needs and the type of albumen. The usual strength is 3 oz. to 1 pt. water; however, for exhibition icing and for use in making the prefabricated designs and stencilled designs, 4 oz. to the pint is recommended. Beating of the icing increases its bulk and creates millions of air cells which spoil the appearance of 'run-outs' and makes the hardened sugar more friable.

For exhibition work and prefabricated pieces, the following method is recommended: Place the strong albumen solution in a clean machine bowl and add to it as much sugar as it will take to make a thick mass. This is usually about 8 lb. Allow the machine to clear this mass on slow speed then add water to bring the icing to required consistency. Keep the icing scraped down by using a wet celluloid scraper. This will dissolve any particles of icing which will form even during the brief time it is exposed to the air in the machine.

Never beat the icing on any other speed but bottom and then only for about 5 minutes. This will give an icing which is ideal for run-out work, etc.

If the sugar is required for piping, then it must be further beaten until it is of the correct consistency. For sugar flowers, practically full peak should be obtained so that the sugar is extremely easy to pipe, being soft and yet able to retain its shape.

For coating, run-out work, and for use with stencils, it is advisable to keep the icing for one or two days so that, when the final conditioning takes place, any remaining air cells are eliminated.

In a dry atmosphere, royal icing dries very hard and, for normal use, a softener is recommended. This can take the form of a hygroscopic substance, such as glycerine, or a jellying agent, such as marshmallow. Rapid hardening is achieved by the addition of a weak organic acid, such as acetic, tartaric, citric acid or lemon juice, which latter imparts a delightful flavour as well. It is important to realize that atmospheric conditions play a very important part in the final result and, even with the use of a hardener, the icing must also have a low humidity to set and dry out.

To achieve good results in run-out work, it is essential to dry them out rapidly and the use of a special drying cabinet is recommended. This also ensures a brilliant gloss which cannot be achieved in any other way.

Softeners, such as glycerine, cannot, of course, be used for run-out work. The use of some acid is recommended but if colour is to be used this must be stable to acids.

Colour may be added in any desired shade and it can be in liquid, paste, or powder forms. Where particularly deep shades are required, it is advisable to use the powder colour so that the consistency is not greatly altered. It should be remembered that colours fade, especially in strong sunlight. Also it tends to migrate to the surface of softened sugar so that matching successive bowls of sugar for colour is virtually impossible. Sufficient icing should be coloured in the first instance to finish the whole cake even if the icing has to be kept for two or three weeks. In such cases it is advisable to store the icing in tins with wax paper sealing the top of the icing and the tin sealed with a tight-fitting lid. For prolonged storage, icing, so packed, then should be placed into a cold room or refrigerator. The sugar icing, for immediate use, should be put into a plastic or earthenware basin and kept covered from exposure to the air with either a plastic cover or a damp cloth. For exhibition work the latter is to be preferred.

Royal icing required for piping must be adjusted in consistency, according to the thickness of the piped line or design. For very fine exhibition piping, the sugar should be relatively soft and the tube placed in a very small bag so that the maximum control may be exercised over the sugar. The use of a small paint-brush is also recommended. Used damp, it will seal the joins of two lines meeting whilst a dry brush easily removes any piped line which is either incorrectly piped or which breaks.

It is difficult to describe the exact consistency required for run-out and stencil work, since this varies not only with the shape of the design but also the nature of the icing. Most beginners make their sugar too soft, with the result that the pieces of work dry out hollow in the centre or fail to dry at all. It is advisable to condition the icing with water to such a consistency that it must be moved before it will flow. This applies to small areas where the icing may be moved with the end of the bag or tube. For larger areas, gentle tapping will achieve a similar result.

Almond Paste

Almost any type of paste can be used in decoration but the two most popular are almond and sugar.

A reasonable almond paste for decorative purposes ought to have an almond content of at least 25 per cent, but this can be reduced further by adding glucose, icing sugar and water or eggs to the manufactured paste. A cheaper medium is sugar paste but this lacks the flavour of good almond paste. It is recommended that the paste should contain a quantity of invert sugar or glucose to keep it mellow and to prevent it drying out.

For modelling purposes, it is essential to use a raw marzipan of $66\frac{2}{3}$ per cent almonds and from this make a paste by using an equal weight of icing sugar, some glucose, and a little water. A small quantity of 'gum traga-canth' may also be added to set the marzipan and to improve its plasticity.

Sugar paste is softer than almond paste and, for certain things, such as modelled fruits, it is difficult to use. It may be easily made by adding icing sugar to marshmallow or pur-chased ready-made in white or in a variety of pastel shades. A sheen may be applied to all pastes by rubbing the surface with the palm of the hand, but care should be taken to ensure that the icing sugar dusting is not done too liberally, otherwise this cannot be achieved.

Chocolate

There are several forms of chocolate which are available as follows:

Couverture – This may be obtained either plain or milk or as white chocolate. In every case it needs to be tempered before use. This imparts a gloss and snap to the set chocolate which enables it to be used in prefabricated off-pieces as decoration. Briefly, the temper-ing process involves melting and heating the chocolate over a water bath or bain-marie to 115° F. for plain, and 110° F. for milk, cool-ing rapidly to setting point (about 80° F.) or thereabouts and gently heating again until about 85° F. All chocolate manufacturers will give tempering instructions with tem-peratures.

The author has used the following method with great success:

The chocolate is broken into pieces and placed into a small bain-marie of hot water at about 120° F. When all the chocolate is thoroughly melted (the temperature is not critical) the pan of chocolate is placed into another bain-marie in which there is ice-water.

Stirring rapidly, the chocolate is quickly brought to the setting point at the base of the pan. When the chocolate has reached a thick consistency, with perhaps particles of solid chocolate mixed in, remove to the hot water again and stir until all the chocolate is melted again. The water at this final stage should be only just sufficiently hot to bring the chocolate to the final condition. After some practice, this may be quickly achieved without the aid of a thermometer, the index finger being sensitive enough to record the correct temperature of the water for each operation.

When the chocolate design has been done, it is left to set in a cool (not cold) room. Extremes of temperature should be avoided. It is just as much a mistake to place chocolate into a cold room or a refrigerator as it is to allow it to set in a warm room.

After tempering, the chocolate should be used at only about 4° F. above its setting point. Spreading the chocolate on to either too cold or too hot a surface or moving it whilst it is setting will cause the chocolate to have the characteristic streaks of sugar and fat bloom.

Baker's Coating – This product lacks the flavour of couverture and many of its quali-ties, such as snap and gloss. Its great advan-tage is that it requires no tempering and one does not have to be so careful in its use. It is not so critical to temperature changes and may be heated to 120° F. The viscosity, how-ever, changes slightly with heat, being thinner when hot than when tepid.

Block Cocoa or Unsweetened Chocolate – This is used for flavouring very sweet icings, especially fondant. It may also be used to flavour creams, cake mixes and marzipan. The unsweetened chocolate is used by first breaking into pieces and melting and then beating or mixing it in. Care should be exer-cised to see that the cream or mixing is not so cold that the chocolate soldifies before it becomes thoroughly incorporated.

Cocoa powder may be used as a decorative dusting medium, but should be mixed firstly with icing sugar to take off the bitter taste.

Piping Chocolate – Melted chocolate is not the easiest material with which to pipe for, unless it is very near to its setting point, it is too fluid, and at this temperature the choco-late may easily set in the tube and bag. Fortu-nately, we can bring chocolate into a piping condition by adding to it water or spirit. Only a few drops of water will alter the consistency of the chocolate and bring it to a condition suitable for piping.

Chocolate so treated, however, loses some of its snap and gloss, although this is reduced by using, instead of water, a liquor or spirit, such as rum. Piping jelly or a gelatine solution may also be added and this tends to overcome the loss of gloss.

Chocolate and Chocolate Flavoured Icings – Besides adding chocolate to icings, a very acceptable icing may be made by adding to melted chocolate, hot sugar syrup. At first the chocolate changes from a fluid state to a thick doughy mass but, as more syrup is added, this changes until a very smooth icing with a brilliant gloss results. It is used hot, like fondant, and sets with a thin crust, whilst the interior of the coating remains soft.

Ganache – This is a cream which should be made from chocolate couverture and fresh dairy cream. There is, however, a number of different recipes for this, some replacing the cream with butter and milk, whilst others use evaporated milk or imitation cream. The true flavour of ganache is not, however, greatly impaired, provided chocolate couverture is used. The usual ratio is 2 lb. chocolate to 1 pt. cream. The cream is brought to the boil and the chocolate stirred in. This may be beaten to a light cream or refrigerated and used as a paste. The consistency of either may be adjusted by the ratio of cream to chocolate.

Various Creams

Buttercream is probably the most widely used cream, although often misnamed, since many are not in fact made with butter but are merely filling or decorating creams. From the point of view of decoration, these creams are all used in similar ways, the only variations being flavour, smoothness, and consistency. Innumerable recipes of these creams may be found in text and recipe books. However, one which has had success in exhibition work is the following German recipe and it is recommended.

Buttercream

Beat to a sponge:
  12 eggs (fresh)
  14 oz. castor sugar

In about four portions add:
  2 lb. unsalted butter
and beat in well.

For the best results the cream should not be put into a refrigerator or suffer extremes of temperatures. Temperature is the main factor in controlling the consistency of buttercreams for piping, although the water or liquid content is also important.

Fudge Icings

Goods which have to be wrapped, are best decorated with a fudge icing or cream which skins and, therefore, has a dry surface. Used hot, these icings may be used like fondant whilst, if left to go cold and then beaten, we have a cream which may be piped.

Nuts and Nut Products

Nuts make an excellent decorating material whether they are whole, split, flaked, nibbed, or ground. They may be used with the natural skin (unblanched) or with the skin removed (blanched). The nibbed and flaked variety may be roasted and the nibbed coloured. The main varieties of nuts used as a décor are almonds, hazelnuts, walnuts, pecan, and cashew.

Coconut is also a popular decorating medium and this can be obtained in strip or desiccated forms. This, too, can be roasted.

All these nuts may be made into pastes and used in a similar manner to almond paste.

Praline

This is a product of nuts and sugar roasted together and milled to a fine, smooth paste. It is a very useful flavouring agent.

Croquant or Nougat

This is another similar product, except that it is not milled to a smooth paste, but may be moulded or cut into shapes or merely crushed and used as granules.

Jams, Jellies, and Conserves

These may be obtained in a variety of flavours and colours and can be used to advantage in the decoration of cakes either on their own or in combination with other decorative materials.

For example, they may be used as a glaze over fruit, or piped into a pattern or used to fill in an area of a pattern made with marzipan, chocolate, or buttercream.

If large areas are required to be smooth, the jam or jelly must be used warm so that it will flow easily.

Sugar and Preserved Fruits

Included in this category are the crystallized fruits as well as the confiture fruits. The only difference between these two is that the latter is made in such a way that the outside sugar layer crystallizes. With crystallized fruits the whole mass has become crystallized.

For decorative purposes, confiture is preferred, as this looks more transparent and has a gloss.

Glacé cherries and angelica are two of the most commonly used fruits, but any fruit may be used with good effect, especially pineapple. A method for preparing fruit confiture is as follows:

Firstly select sound ripe fruit. Prepare a sugar syrup made with cane sugar and bring to a density of 18° Beaume, using a saccarometer. The fruit is prepared by peeling if necessary and cutting it into rings or cubes or leaving it whole. These are now placed carefully into the syrup and the whole brought to the boil. The fruit is then removed from the heat and left for 24 hours. After this time, the fruit is carefully placed on a draining wire.

More sugar is added to the syrup to increase the density by 2° B. so that the saccarometer reading is 20° B. The fruit is replaced and the whole brought to the boil when again it is removed from the heat and left for 24 hours. This process is repeated on successive days, increasing the density by 2° B. daily until 28° B. is reached. Then, instead of adding cane sugar to increase the density, glucose is added. When the density is 32° B., the fruit is removed and drained, after which it is ready for use.

Confiture Pineapple

A quicker and more convenient method of preparing confiture from tinned pineapple is as follows:

Recipe

One A10 tin pineapple (approx. weight $6\frac{3}{4}$ lb.)

3 lb. sugar

$\frac{1}{4}$ oz. citric acid

4 pints water (including juice of fruit)

Method

1 – Pour off the juice of the tinned fruit into a measure and make up to $\frac{1}{2}$ gallon with water.

2 – Add the sugar and acid and bring to a clear syrup with heat.

3 – Immerse the pineapple and bring to the boil.

4 – Remove from the heat and leave for 24 hours.

5 – On the second day, replace over the heat and re-boil.

6 – Remove from the heat and leave for 24 hours.

7 – Repeat operations 5 and 6 for five more days, making seven days intermittent boiling in all.

8 – At the end of the process, remove and store for future use.

NOTE: The pineapple may be left in the syrup until required. The syrup itself may be used as stock syrup for reducing fondant.

Glacé fruit, such as cherries, may be used individually; but most fruits, such as pineapple, angelica, etc., will need to be cut into pieces of definite shape and size. Sometimes a good effect may be obtained by chopping the fruit finely and massing it in certain areas of the design.

Tinned fruits are particularly good for some forms of decoration on cakes, particularly with quick-selling lines such as torten. To obtain the best effect from their use, they should be glazed with a puree, jelly, or syrup.

Fresh and frozen fruits have a limited use as a decorating medium because they deteriorate so quickly. If they are used they should be covered or glazed with jelly or puree.

Jelly Decorations

There are many jelly decorations which may be purchased and can be used to good effect. These include orange and lemon slices, pineapple rings, strawberries, and raspberries.

Crystallized Flowers

Rose, lilac, violet, and mimosa flowers can be obtained in a sugar-preserved form and used as very effective decorations either on their own or in conjunction with other materials. The yellow mimosa balls, for example, form very good flower centres.

Sugar

It is not always fully appreciated that the various forms of sugar can also be used in various aspects of decoration. Icing sugar, castor, granulated, nib, and demarara all have their own particular uses. For example, icing sugar on a chocolate log not only gives a good snow effect but adds to its attraction.

Crumbs

Other materials, which are not always exploited as a decorating medium, are cake, or jap crumbs. These may be roasted and used either for masking the sides of cakes or for dredging over different areas of the design, such as the centre of a torten.

Chapter II

Prefabricated Decorations (1)

IN recent years, there has been a greater tendency for more prefabricated decorations to be purchased from allied traders and used for the decoration of cakes. This has become necessary because of the lack and cost of skilled labour.

Much of this prefabricated decoration is expensive to buy, and even decoration, such as cherries and nuts, can cost far more than that decoration which can be made by the confectioner.

Much of the décor illustrated and described in this chapter can be produced by unskilled labour, once a small initial period of training is given. Such décor may be done at slack periods and stored for future use. Apart from the saving in cost of materials, the use of the infinite variety of décor which can be made, is likely to have the effect of increasing sales. Another saving is in the labour of applying such décor to a cake.

How much easier and quicker to use a dry décor, such as a chocolate or sugar button. As can be seen from the table of comparative costs, it is also so much cheaper.

Table of Comparative Costs

Décor	Number per oz.	Cost per oz. (pence)	Approx. cost each (pence)	Remarks
Alpine cherries	12	4·875	0·406	Average size
Walnut halves	16	5·68	0·355	Average size
Almonds	23	6·125	0·266	Whole unblanched
Hazelnuts	36	5·25	0·328	Whole unblanched
Chocolate buttons	20	5·00	0·25	Chocolate couverture
Sugar buttons	32	1·25	0·039	Royal icing
Almond paste cut-outs	36	2·93	0·81	Approx. 1 in. square almond paste
Chocolate cut-outs	36	5·00	0·138	Approx. 1 in. square couverture
Almond paste flowers	42	2·93	0·0697	Without jelly filling

The above décor has been carefully selected for a uniform size so that costs are truly comparative.

* These prices are as from November 1977.

Sugar Balls

Fig. 5 – These are made by piping small drops of softened coloured royal icing into a tray of flattened castor or granulated sugar. These are then completely covered in the same sugar and left overnight to dry.

Afterwards they can be combed or sieved out and stored in jars in a dry place. The amount of raw material used for these is negligible: it is the operative's time which is the largest cost involved. However, these are so quickly done that it becomes a very economical décor to use. One very great advantage is that the size and colour can be varied to suit individual requirements.

No bought decorations can imitate these sugar balls either in size or colour. The only other comparable decoration is sugar mimosa balls which are larger and, of course, only suitable in yellow.

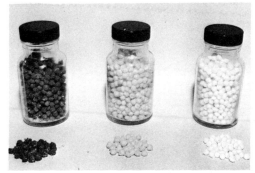

Fig. 5

Sugar and Chocolate Buttons

Fig. 6 – These offer us a very large assorted range of décor because, firstly, the colour can be varied from white through all the shades to chocolate and, secondly, there are so many materials with which they can be dressed, as the following list demonstrates:

Castor sugar	Nib almonds
Granulated sugar	Small flaked almonds
Demarara sugar	Toasted nib almonds
Nib sugar	Toasted flaked
Sugar vermicelli	almonds
Chocolate vermicilli	Toasted desiccated
(milk and plain)	coconut
	Toasted desiccated
	coconut

This makes twelve types of dressings which, if permutated with a range of colours, result in a very large possible variety.

The buttons are made by piping out the sugar or chocolate on to wax paper and sprinkling on the appropriate dressing. In the case of sugar buttons, the excess dressing can be removed by shaking it off the sheet or with an airspray. With chocolate, however, the medium is so fluid, that it is advisable to delay this until the chocolate is firmly set. A useful variety of button may be made by dusting icing sugar on to the button after it is made, as illustrated. Yellow sugar buttons, dressed in castor or granulated sugar, may be especially used to advantage for flower centres.

12

Fig. 6 (Enlargements show actual size)

13

Chocolate Décor

Fig. 7 – In addition to the buttons just described chocolate may be used for a variety of prefabricated décor.

Chocolate Flowers – Flowers are easily made by piping out five or six dots in the form of a rosette and placing a sugar ball, preferably of a yellow colour, in the centre. A large silver dragee could also be used for this purpose with good effect.

Filigree chocolate flowers may also be made by just piping the outline in chocolate and placing a yellow ball in the centre. (*Top left*.)

Chocolate Cut-outs – For chocolate cut-outs the chocolate is firstly tempered (see Chapter I) then spread out on to a sheet of greaseproof or wax paper. Just prior to setting, the chocolate is cut into the desired shapes either with a knife or cutter. When completely set, the shaped pieces are removed from the paper and used, whilst the cuttings are put aside for re-melting and tempering.

The cut-outs also lend themselves to further treatment and may be dressed with any of the materials already listed.

This dressing must be done, of course, before the chocolate has set and is cut into shape. (*Top right and middle left*.)

Chocolate Curls – There is a number of ways by which chocolate may be made into curls. The author has used the following method with success: it is, therefore, recommended.

Pour chocolate out on to a marble slab and spread it thinly to and fro with a palette knife until the chocolate begins to set. Then, before it hardens, completely cut off the curls by using a long knife in a shearing action and at an acute angle to the marble. The size of the curl is governed by the amount of chocolate which is cut off in the shearing action. The length of the curl may be determined afterwards by cutting with a hot knife. (*Middle right*.)

Chocolate Run-outs – Being a semi-fluid medium, chocolate may be run out to any desired shape to suit a particular purpose. This shape may be firstly drawn out on paper and used in the way illustrated. A sheet of wax paper is laid over the drawing and the chocolate is piped on to the outline which shows through. Such run-outs can also be dressed and this, too, is illustrated, using chocolate vermicelli as the dressing medium. Almost any shape or pattern can be reproduced in this way and used for decoration of fancies, gateaux, or torten. (*Bottom*.)

Fig. 7

15

Christmas Logs

Fig. 8 – These are easily and quickly produced and are a very effective decoration for Christmas cakes and gateaux.

Ropes of almond paste having been made, they are covered evenly with chocolate. Just before the chocolate is set, it is combed with a ribbed scraper to give the realistic log bark effect. The ropes are cut into lengths of approx. 1½ in. and are then ready for use. For more effect, the logs can be dusted with icing sugar to give the impression of snow.

Fig. 8

16

Snowballs

Fig. 9 – These are easily made by rolling pieces of almond or sugar paste into balls. The eating quality of such décor, however, leaves much to be desired and the use of cherries as a base is to be greatly preferred.

The cherry must be completely covered in a white medium before it becomes acceptable as a snowball. White sugar paste is a very suitable material; alternatively, the cherry may be enrobed in a hot, thick fondant.

Afterwards, the cherry balls may be either rolled in icing sugar or royal icing to give them a rough-cast effect, as illustrated.

Fig. 9

Almond and Sugar Paste Cut-outs

Fig. 10 – Cut-outs from almond or sugar paste make an effective and quickly used décor for the finishes of fancies, gateaux and torten. Although the variety of possible shapes is vast, we are limited to those shapes which can be cut out either with a knife or cutter. A number of such shapes are illustrated here. However, for speed of production the square or rectangle has a great advantage. These may be cut out very speedily by means of the caramel cutter as shown.

Like the buttons and chocolate cut-outs, these, too, lend themselves to a similar treatment. They may be coated with a medium, such as chocolate, fondant, jam, or royal icing and either left in this medium, or further dressed with any of the materials previously listed. Even the coating medium may be textured and, in the illustration, we see squares on which the royal icing used is rough cast; sprinkled with crushed crystallized flower petals; dressed with toasted flake almonds: and dredged with castor sugar.

Paste also lends itself to treatment by special rollers which leave attractive impressions. Four examples of these are also shown.

Another effective décor can be applied by spinning chocolate over the paste. The technique of this is depicted, as well as the effect on both ribbed and plain squares.

Of course, chocolate may be spread on to the paste in the same way as royal icing and either dressed with another material or textured. A comb scraper used to put on a ribbed effect looks very good. In this case the chocolate should be slightly thickened by adding a drop of water beforehand.

Although emphasis has been given to further dressings of these paste squares, they may be used to good effect on their own, if suitably coloured and fine castor sugar used as the dusting medium. The sugar is then rolled into the surface of the paste: it provides an attractive crystallized appearance which is far superior to a plain matt or even glossy surface.

Fig. 10 – Top left, marzipan cut-outs; top right, spinning chocolate over ribbed marzipan, prior to cutting; middle left, textured marzipan cut-outs; right, use of roller; bottom, examples of dressed cut-outs.

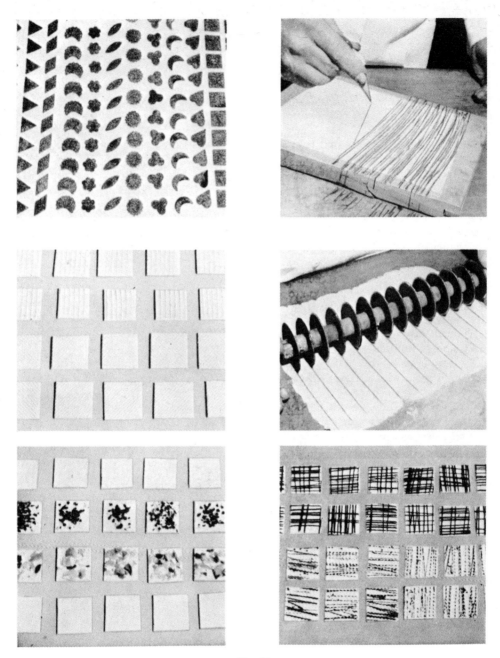

Fig. 10

19

Flowers

Fig. 11 – The flower shape is easily cut out of paste with an appropriate cutter. They may be made from either plain or crystallized paste and be embellished with a variety of different centres, many of which are described under the heading 'Sugar Buttons'.

A most attractive flower may be made by cutting the petal shape out of thinly pinned out paste which has been 'crystallized'. The flower shapes are laid in rows on to grease-proof paper and cupped by pressing into the centre, the rounded end of a piece of stick about $\frac{3}{8}$ in. thick. The flower shapes are then dried and can be tiered in storage until wanted. Before being used, the centre is filled with coloured fondant or jelly. This should be first coloured a bright yellow and piped in a heated condition so that it flows and sets with a slight skin, which makes it easier for handling.

A shape similar to a flower may be made as shown in the last illustration of the figure. The paste pieces are merely folded over and jelly piped into the cavity so formed.

Fig. 11

20

Other Cut-out Shapes

Fig. 12 – Almost any shape may be cut out of paste provided a suitable cutter is available. Here illustrated are three shapes which can be used to good effect at Christmastide.

The Christmas trees are cut out of green almond paste, and placed in rows on to dampened greaseproof paper or parchment paper. It is important to secure these cut-outs to damp paper so that when they are sprayed the air pressure will not blow them away or out of position. Also it makes for easier handling afterwards. They are then sprayed with the air-brush to give it more interest, the colour used being a mixture of green and chocolate.

Candles are similarly cut out, except that white paste is used. The only other treatment required is a touch of colour from the air-brush, to imitate the flame.

The holly leaves receive the same type of treatment as the Christmas trees. When set and dry, however, a bulb of red, royal icing is piped on, to give the impression of the berry. This is done in softened icing, which has been suitably coloured a bright red, using powdered colour. When all the berries have been piped on, the whole sheet is subjected to top heat for a few minutes. This increases the gloss of the berry and makes it look more realistic.

All these cut-outs may be sprayed over with an edible varnish and stored for a long time before use. The varnish not only helps to keep the paste free from damage by moisture but also discourages mites which feed on such foodstuffs.

The design of the cutter is worthy of consideration. It should be made so that it tapers away from the cutting end. By using this type, several pieces can be cut out before having to be removed. This is then easily effected by turning the cutter upside down when the pieces will fall out in a neat pile.

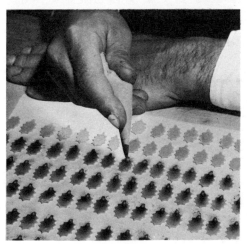

Fig. 12

Chapter III

Prefabricated Decorations (2) – Stencils*

Stencils

THE use of stencils as an aid to the rapid repetition of decorations is hereby exploited.

It is assumed that most readers will understand what is meant by a stencil. However, there is a number of materials from which these can be made and some comments on this might be helpful.

Oiled Parchment (known as stencil paper)

This material is by far the easiest to cut, is cheapest to use, and is very flexible. The design has to be traced in pencil then cut out with a craft tool. It has one serious disadvantage in that it will not stand up to rough usage and, if folded, will crack and break at the crease. Nevertheless, in the hands of a careful operator, stencils made with this paper have a very long life.

Although this material is paper, the fact that it is oiled will allow it to be placed into cold water, and thus it can be easily washed free of any sugar after use. After drying off the surplus water with a dry cloth or paper towel, the stencil is best stored between sheets of greaseproof paper on a flat tray. Most of the stencil designs illustrated in this book have been executed in this material.

Treated Cardboard

This is cardboard which has been treated with French or button polish after the impression has been cut (see page 219).

---

* Details of stencil designing are given in the book *Cake Design and Decoration*, by Hanneman and Marshall.

Plastic Film

This is a transparent film which is obtainable in several grades. Only the thin grade is flexible enough for stencils. This material can be cut with a sharp craft tool to a pencilled design to which it is firmly attached and which can readily be seen through it. Great care must be taken in the cutting, however, since the material easily rips, and this can be a serious disadvantage.

Its advantages are its transparency, which allows it to be placed accurately into position on a cake, and flexibility, by which it can be applied to any surface, round or irregular.

Being impervious, it can be left in water for the sugar to soak off after being used and it can be left to dry without being wiped with a cloth. It is also fairly cheap to buy in the roll.

Although superior it is difficult to cut successfully.

Celluloid

This is also difficult to cut and does not have the same flexibility as the film. It is, however, a very much stronger material and, provided it is not brought into contact with heat, which distorts its shape, or flame in which it ignites, it should have a much longer life and stand up to more rough usage.

Sheet Copper

For durability, this is by far the most successful of all materials from which stencils can be cut. The only disadvantage is the cost of getting the stencil made, since this is a difficult task to do by hand and is best put into the hands of a professional engraver. However, since the cost is only likely to occur once and is easily off-set against the saving of labour in its use, this can be justified.

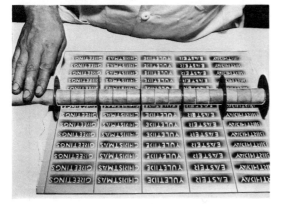

Inscriptions

Fig. 13 – Various inscriptions may be stencilled out on almond or sugar paste plaques. In the illustrations we see how five of these may be done simultaneously, stencilled in rows of 14, making 70 such plaques manufactured in one unit operation.

Almond or sugar paste is pinned out to about $\frac{1}{16}$ in., a mask is laid on and chocolate colour sprayed through, so that it is more concentrated at the centre of the rectangle. When dry, the stencil mat, containing the inscriptions, is laid over and white royal icing is spread through. The mat is now carefully removed to reveal the inscriptions neatly stencilled on to the chocolate-coloured background.

The inscriptions are separated by using two cutters, one with the rotary cutting discs set closer together than the other. One set is rolled across to separate the 14 words and the other to divide the five rows.

There are several options to this treatment. Instead of colouring the paste using the airbrush, chocolate or another suitably coloured paste could be used. Alternatively, a darker coloured icing could be stencilled on to a pale-coloured base. The main aim is to provide a contrast between the inscription and its background.

Fig. 13

23

Flower Petals and Leaves

Fig. 14 – Although the stencil material must be very thin, this does not limit the amount of sugar which may be spread over to fill the impressions. If a semi-soft icing is used and spread fairly thickly over the stencil, forms may be stencilled, which have exactly the same characteristics as run-outs. Care must be taken, however, when removing the stencil not to smudge or distort the shape. The consistency of the sugar should be such that it only just flows and, in fact, may require gentle vibration or tapping to enable it to do so. Further details about the sugar required for run-out work will be found in Chapter I.

Petal and leaf shapes lend themselves extremely well to repetition by this technique. Royal icing, of an appropriate colour, is spread over the stencil mat and deposited as the required shape on to a sheet of wax paper. The board on which this is placed is tapped to eliminate the rough edges of the shapes and is then put into a drying cabinet. If desired, the shapes may be dredged with castor sugar to give them a crystallized appearance. Further, the shapes can be tinted any desired colour and shaded with the air-brush to make them look more attractive. This is done by directing the air-brush at one edge only and moving it along in one direction. The holly leaves, on which this treatment is illustrated, may be finished off with a red piped berry as in **Fig. 12**, or assembled on the cake as a spray before the berries are applied.

The heart illustrated is a versatile shape which may be used for flowers, petals, leaves, or a decoration in its own right. For special occasions, such as Valentine's or Mother's Day, the heart-shape may be used to advantage in all types of gateaux, fancies, and torten.

Fig. 14

Flowers
Fig. 15 – This technique may also be applied to certain flower forms and three are illustrated. The yellow centres can be placed into the centre of the stencilled shape as soon as it is done, so that it adheres firmly. The small flower shapes have centres of small, yellow, sugar-piped balls. Mimosa balls and large flower-centres have been used for the large flowers.

Fig. 15

26

Strawberries and Chicks

Fig. 16 – There are many other forms which may be treated in the same way. Fruits are very suitable for this treatment and, in the illustration, we see two types of stencilled strawberries. One type has been sugar-dredged whilst the other has been spotted with a stiff nylon brush dipped in red colour as shown. Both are finished off with a calyx piped in green sugar.

Chicks may also be executed in the same way. These, too, are sugar-dredged. When dry, the eyes are put in with a brush or pen dipped into chocolate colour. The beak may either be painted with an orange colour or sprayed on using the air-brush. Because of the granular surface, this last method gives the best results.

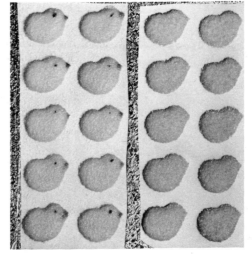

Fig. 16

27

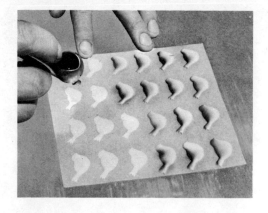

Robins

Fig. 17 – For Christmas cakes and gateaux, the robin shape may be used to very good effect. Stencilling enables us to produce these speedily in large numbers. Firstly, the shape is stencilled out in white royal icing and thoroughly dried off in a drying cabinet. Chocolate colour is then sprayed on the top and underside of the tail, leaving the front still white. A patch of orange-red is applied with the air-brush to this area so that the underside is still left white.

Finally, the eye and beak are applied with chocolate colour from a pen.

The tray of completed robins may now be stored away ready for use.

When these robins are used for a cake decoration, the legs and feet have to be piped with a No. 1 tube.

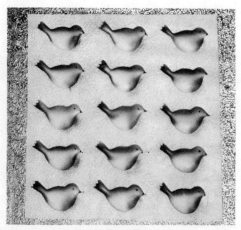

Fig. 17

28

Butterflies

Fig. 18 – Butterfly wings may be quickly executed by stencilling the wing forms in pairs and in separate rows. When dry, these are then sprayed with colour from an airbrush.

The colours recommended for this are pale green for where the wing joins the body and orange for the wing-tips. The bodies are piped in with softened, coffee-coloured icing, as illustrated. These are picked up and placed into a Vee-shaped template in which they are allowed to set. Chocolate spots may be applied to the wings with a paint-brush if desired. The colour combinations may be varied with very good effect.

The wings, for example, could be mottled with either the paint- or air-brush. To use these decorations, they can be easily removed from the wax paper and placed on to the cake.

These techniques remove the execution of run-outs from the restricted field of exhibition work and make it a sound commercial proposition.

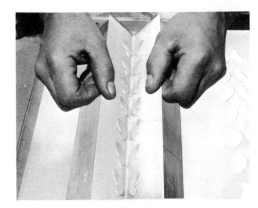

Fig. 18

29

Chapter IV

Fancies (1)

MANY designs and ideas are provided in this chapter for the decoration of fancies. Some of them will be original, whilst others will have been used before but are such good basic designs that they are included.

Although many of these designs could be used for cheap, commercial fancies, some require to be executed by a skilled decorator and are, therefore, more applicable to high-class fancies and exhibition work.

For convenience, all the designs have been executed on square bases, but most of them could be used for any shape. The confectioner will make his selection according to purpose.

Basic Shapes
Fig. 19 – Here we see twelve basic shapes for genoese fancies. However, only four of these shapes are strictly commercial. These are those which can be cut from a sheet of genoese with the minimum amount of trimmings and cuttings, and are the square, triangle, diamond, and rectangle. All the other shapes require special cutters and, besides involving much time, a considerable amount of scrap genoese results.

The author has often seen cases where a confectioner cuts out shapes for his fondant dips, practically wastes his scrap genoese and then fails to spend sufficient time on decorating, saying that he cannot afford the time on a cheap fancy. This is a false economy, because if a fancy is to be sold cheaply, there is no sense in using fancy shapes. In any case, the extra time spent in cutting out the shapes could be more usefully spent on the decoration – even the selling price of the fancies might be increased as a result!

Naturally, some designs lend themselves to one shape better than another, but the reader will be able to judge this for himself, if the designs are closely studied.

Fig. 19

Basic Designs

Fig. 20 – These basic designs are used in several ways and in the following figures they are reproduced on to actual fancies:

Fig. 20

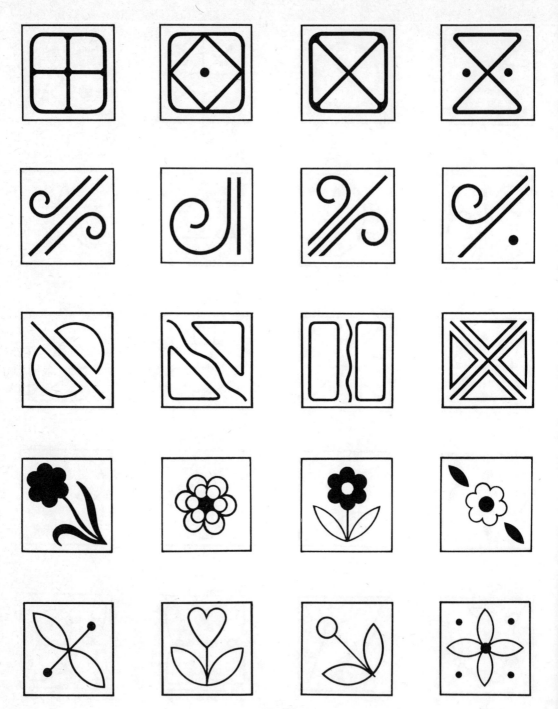

Fig. 20

34

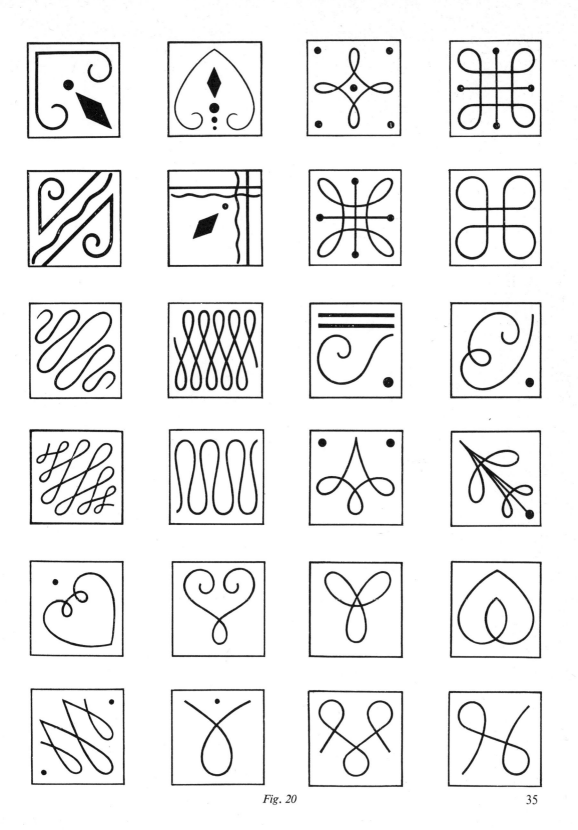

Fig. 20 35

Figs. 21, 22 and 23 – The treatments given to the designs illustrated fall into three categories as follows:

1 – The design is executed on to the surface of the fancy and just embellished with sugar balls.

2 – After the design is piped, jelly of the appropriate colour is run into some of the areas enclosed by the piped line. This jelly should be heated so that it easily flows into the area. This will also ensure that a skin is formed that will facilitate packing.

3 – The design is reproduced in chocolate and this when set is transferred to the fancy.

Fig. 21

37

Fig. 22

Fig. 23

Chocolate and Almond Paste Décor

Fig. 24 – Crystallized almond paste and chocolate cut-outs are used in many of these illustrations. Notice how these may be combined with each other and jelly to give variety. The use of chocolate pieces, which have been dusted with icing sugar, is also seen to advantage on chocolate-coated fancies. Also seen here are the uses of chocolate curls, almond or sugar paste flowers, and filigree chocolate flowers. The making of all this prefabricated décor is fully described in Chapter II.

FLAVOUR

Before pursuing more ideas for the decoration of fancies, some details on how flavour may be created might be helpful. Flavour may be incorporated into a fancy in the following ways:

1 – The genoese itself can be flavoured with (a) essence or essential oil, (b) fruit concentrate, or (c) addition of crushed or chopped fruit or nuts.

2 – The genoese can be layered with (a) a fruit curd or preserve, (b) a cream containing either flavour or the crushed/chopped fruit or nut, or (c) the crushed/chopped fruit or nut itself.

3 – If the genoese is to be enrobed with fondant or another type of icing, this can be suitably flavoured with either natural or artificial flavours. The use of special sugar makes it possible now to form a fondant from the pure juice of a fruit, which is delicious.

4 – A suitably flavoured cream, jelly, or almond paste piece, can be placed on to the genoese and then enrobed with the fondant.

5 – The fruit itself can be (a) placed on to the genoese and enrobed, or (b) placed on top of the finished dipped fancy.

Fig. 24

Fruit Fancies

Figs. 25 and 26 – The flavours illustrated in **Fig. 25** are apple, orange, peach, pineapple, strawberry, and cherry, and each illustrate four stages of the following technique:

1 – A suitably flavoured cream is piped on to the almond paste covered genoese, in a shape appropriate to the fruit to be depicted.

2 – This is then enrobed with fondant.

3 – The fruit shape is sprayed with an appropriate colour, *i.e.* red for cherry, with the air-brush.

4 – Finally, the finish is applied, using angelica and sugar leaves or any other kind of treatment which will make the effect more realistic.

Although not shown here, many other fruits could be imitated in the same way, such as lemon, pear, and plum.

The fancies illustrated in **Figure 26** have had a slightly different treatment in that, instead of cream, the actual fruit or jelly imitation has been used. We see here four stages leading to two varieties of finish for each fruit. These are as follows:

1 – The glacé fruit or jelly slice is laid on to the almond pasted top.

2 – Fondant is used to enrobe the fancy.

3 – The fruit shape is sprayed an appropriate colour.

4 – The same fruit or jelly slice is used after the fancy is enrobed.

The fruits used for these examples are cherries, orange jelly slices, crystallized pineapple and dates.

Fig. 26

42

Fig. 25

43

Christmas Fancies

Fig. 27 – For a small Christmas decoration, the holly motif is ideal. Besides the use of the prefabricated paste and stencilled holly shapes shown, there are another four ways by which this decoration can be accomplished. These are shown in the first photograph of the figure and are described thus:

1 – *Top left*. Here the outline is first piped in chocolate and the shape infilled with warm green piping jelly. Two bulbs of bright red jelly are piped on to represent the berries.

2 – *Top right*. The treatment here is the same except that no jelly is used, the colour being first sprayed on through a stencil and then outlined in chocolate.

3 – *Bottom left*. Green royal icing has been used here to pipe the outline of the holly and red jelly for the berries.

4 – *Bottom right*. In this example the holly motif is stencilled straight on to the fancy in green-coloured royal icing. This can be further sprayed if required. Berries could be of either royal icing or jelly.

For commercial fancies, the use of a prefabricated holly leaf with a simple line design in chocolate or fondant is recommended. The green of the holly shows up extremely well if pink is chosen as the base colour.

The cherry snowballs and other cut-out motifs used in the examples illustrated are fully described in Chapter II.

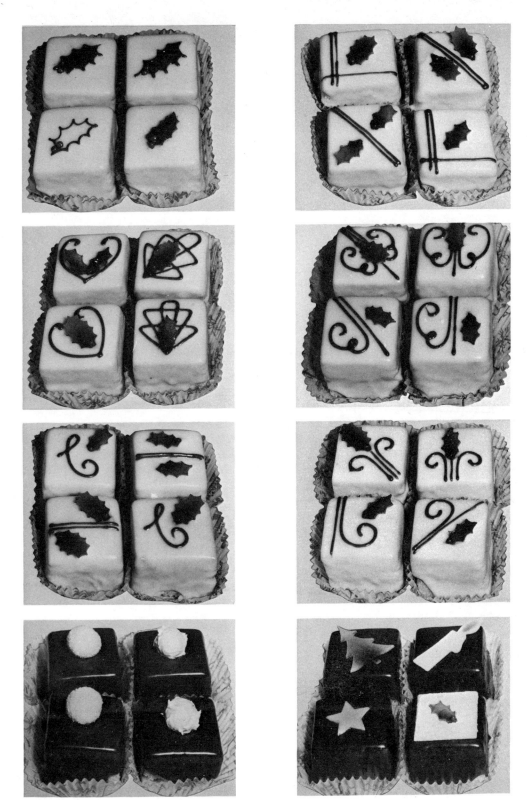

Fig. 27

Chapter V

Fancies (2) – Stencilled Tops

STENCILLED almond or sugar paste makes an ideal top decoration for a fancy. It is easily applied and provides a perfectly dry decorative top which can be wrapped without risk of adherence to the wrapping.

The shape chosen in the following illustrations is again in the square. However, by angling the direction of the cutters, diamond, triangular, and rectangular shapes can be cut just as easily, thus providing a wide variety of appealing fancies.

Almond, sugar, or coconut-paste can be used as a base on which the sugar design is stencilled and this technique is described and illustrated in the figure following.

Stencilled Squares

Fig. 28 – Firstly, the paste is pinned out to about $\frac{1}{16}$ in., the stencil mat laid on, and the paste trimmed to the same size. Royal icing of the desired colour and consistency is spread through the stencil on to the paste underneath with a spreader of wood, plastic, or metal. The mat is carefully lifted away to reveal the chick shape which has been deposited on to the paste underneath. Whilst still wet, these shapes are dusted with castor sugar and the surplus sugar removed by blowing off with the air-brush lead. The sheet of paste is now cut into squares with the caramel cutter which is rolled over first in one direction and then again at right-angles between the stencilled impressions. Theoretically, any number of squares can be cut in the operations; the only limiting factor is the length of the cutter and its number of blades. Obviously the design of the stencil must conform to the spacing of the cutter blades so that the chick impressions are accurately positioned in the centre of the square.

The beaks are tinted with orange colour either by using an air- or paint-brush. The eyes are piped in with softened chocolate icing or marked with a pen-nib dipped into chocolate colour. This latter method is quicker and is preferred for unsugared impressions. With the sugared varieties, however, it tends to spread into a patch like ink on blotting-paper.

The following illustrations of fancies are decorated with stencilled squares produced in this way. Only the actual design of the stencil differs.

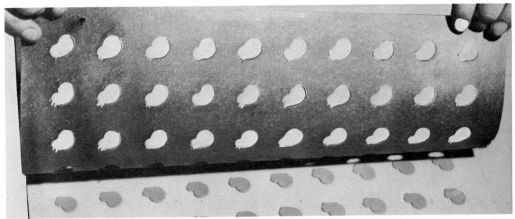

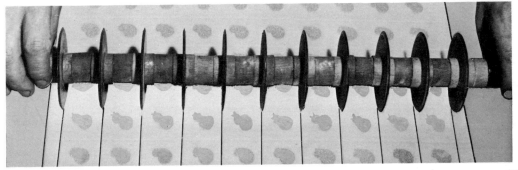

Fig. 28

Two-colour Stencils

Fig. 29 – Where there are two basic colours in a design, such as the red of a flower and the green of its foliage, it is of great advantage to use two stencils cut so that the colours may be stencilled separately. This is illustrated as follows:

Firstly, a green stem and leaf are stencilled on to the paste and left to dry: this should only take about half-an-hour in a warm bakery. Another mat, containing the flower impression, is then laid over in exactly the same place and white icing is spread through. This is dusted with castor sugar so that, when the mat is removed, the flower impression is crystallized whilst the leaf is glossy. The castor sugar dressing is, of course, optional but the contrast of texture looks more pleasing. All that now remains is for the paste to be cut into squares and a yellow icing sugar bulb piped into the centre of each flower. The designing and cutting of these two stencils must be done with care to ensure that the flower is positioned accurately in relation to the stem and leaves. Accurate positioning of the stencil is also important but, provided both stencils are of equal dimensions and the paste trimmed to the same size, this is not difficult to achieve.

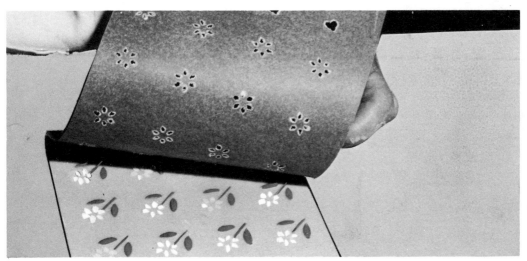

Fig. 29

Easter Fancies

Figs. 30 and 31 – The fancies illustrated in **Fig. 30** are described as follows:

Top left – The chicks are stencilled in bright yellow icing which are sugar-dredged.

Top right and *centre left* – In the two sets of rabbits, chocolate icing was used and half of these were dusted with castor sugar.

Centre right – Symbolic of Easter is The Cross. This is here executed in chocolate icing.

Bottom – The two-colour, stencilled squares have been used for the last two sets in this figure. Four variations of the flower are shown and these are: Petals and centre left, unsugared; both petals and centre, sugared; centre only, sugared; and only petals sugared.

Fig. 30

In **Fig. 31** we show more variations of this type of treatment:

Top left – The chick, emerging from its shell, is treated in two ways: by stencilling in two colours, white for the shell, and yellow for the chick, and

Bottom left – Stencilling in one colour and afterwards colouring with a paint-brush. Three of the chicks shown have received a sugar dressing.

Catkins, too, are symbolic of Easter and its attractive form of design is very popular. In the sets illustrated, this design is executed in four different ways.

Top left – Two stencils have been used, one for the branch done in chocolate and one for the flower executed in white.

Top right – This has been stencilled in the same way except that the white flower is dusted with castor sugar.

Bottom left – The branch is stencilled, but the flower is piped on in white royal icing.

Bottom right – Same as previous one, except that the flower is dusted with castor sugar.

These piped flowers appear superior to the other forms because they are so bold. However, for speed of execution, the use of two stencils is to be recommended.

The large flower form shown also has four variations.

Top left – Red stencilled petals with yellow stencilled centre dusted with castor sugar.

Top right – Red stencilled petals with piped yellow centrepiece.

Bottom left – Chocolate stencilled petals with piped yellow sugar centrepiece.

Bottom right – White stencilled petals with stencilled yellow centre dusted with castor sugar.

The remaining sets in this figure show the use of prefabricated decoration already described in Chapter II.

Fig. 31

53

Christmas Fancies

Fig. 32 – Christmas motifs have been used as a stencil decoration in the eight fancies illustrated. These are briefly described:

Holly and Christmas Trees – These are stencilled in green royal icing and tinted afterwards with a darkened green from an airbrush. If desired, the tubs of the trees may be touched up with chocolate colour. The holly leaves are finished with a small bulb of red sugar piped at the base to imitate the berry. These squares should be subjected to top heat for a few minutes immediately afterwards to set the gloss.

Mistletoe – The mistletoe is stencilled in pale apple green with the stem painted chocolate. This form and that of the tree would benefit from the technique already described, using two stencils so that two colours may be stencilled. In this way the tree tubs and the mistletoe stems could both be stencilled in chocolate-coloured icing.

Stars, Bells, Christmas Tree Branches, and Lantern – All stencilled in chocolate icing. An alternative treatment for this lantern is to stencil the outline only on to the base, to which has been applied orange colour from the air-brush. This would give a much more realistic effect.

Candles – These may be stencilled simply in white and the flame touched up with orange colour afterwards.

Fig. 32

Fruit Fancies

Figs. 33 and 34 – Many fruit forms lend themselves to this stencilling technique and here illustrated are twelve forms:

Fig. 33 – **The orange** has been stencilled in orange-coloured sugar and docked with a stiff nylon brush to give it the impression of the roughened skin.

The lemon is similarly executed except that yellow sugar is used and a touch of green with the air-brush is applied to the top.

Apples have colour applied with the air-brush to a pale yellow base. The colours used are green and red.

Pear. The base colour is the same and red and chocolate colour is used in the brush. If desired, light chocolate spots can be applied to the pear with a stiff nylon brush afterwards. The stalks may be stencilled in chocolate icing, piped, or left out entirely for both pears and apples.

Bananas are stencilled in yellow sugar and marked in chocolate with a paint-brush.

Plums may be stencilled either in yellow or red with the stalk piped or stencilled in chocolate or omitted entirely

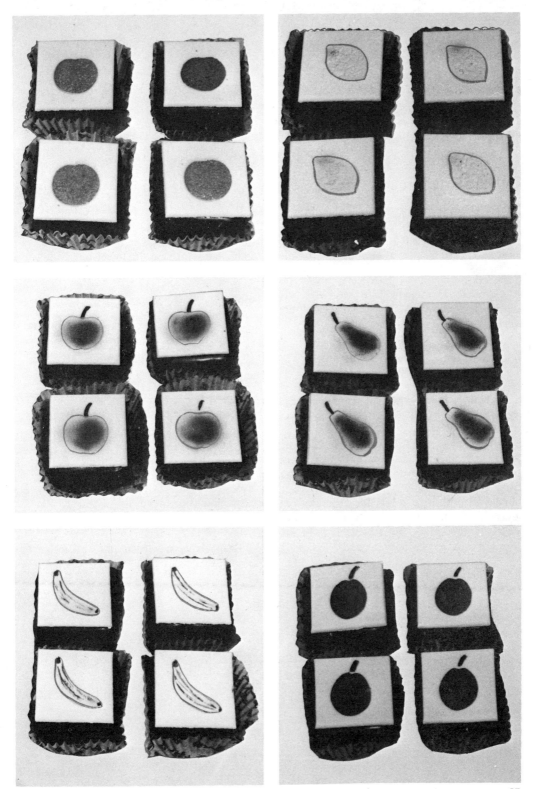

Fig. 33

Fruit Fancies – *continued*

Fig. 34 – For the following fruit designs it would be of great advantage to use two stencils, one for the fruit and another for the stem and leaves:

Cherries are stencilled in red whilst the stalks are executed in chocolate-coloured icing.

Grapes may be done similarly in pale green.

Peach is done in a similar way, the leaves and stem is green, and the fruit in a peach colour sprayed pink with the air-brush.

In all the examples shown, in which the air-brush is employed, colour was sprayed on to the fruit without any masking out of the marzipan. This necessitates complete control of the air-brush and it is advisable, therefore, to mask out the marzipan by re-laying the stencil mat over the design, if the brush is to be used by an inexperienced operative.

Pineapple has two different treatments. On the top two fancies the green leaves are stencilled whilst on the lower ones, the leaves are piped. The stiff nylon brush has again been used to effect a roughened surface and spots of yellow brown mottling.

Strawberries can be treated in a number of different ways as follows:

Top left – Fruit dusted with castor sugar and calyx stencilled green.

Top right – Fruit unsugared with a different type of stencilled calyx.

Bottom left – Similar to the sugared variety except that it is left plain and stippled with the stiff nylon brush dipped into red colour.

Bottom right – This shows another type of calyx, this time piped.

(Strawberry red colour is used in each case.)

Raspberries are executed in a similar way.

Fig. 34

Run-out Fruits

Fig. 35 – Many of these fruits, such as strawberries, raspberries, pineapple, plums, bananas, lemons, apples, pears, and oranges, can be stencilled as run-outs, colour sprayed and, when set, placed directly on to the fancy.

Here we see two treatments of a strawberry used in this way. The two top fancies have fruits which are sugar dressed, whilst the two lower fancies are decorated with run-outs, which have been stippled with red colour. The calyx in every case was piped in.

Direct Stencilling

Fig. 36 – If desired, the designs in this chapter may be stencilled directly on to the fondant-covered fancy as the figure illustrates. Naturally, the process is much slower because each fancy has to be done individually but the effect is just as pleasing.

Stencilled Run-out Flower Forms

Fig. 37 – Here we show how run-out flowers and leaves may be used to decorate both white and chocolate fondant covered fancies. To create contrast, the white fancies were decorated with a red flower, and the chocolate with a white one.

Fig. 35

Fig. 36

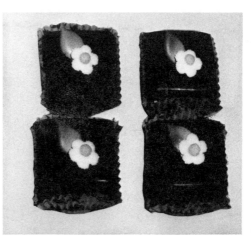

Fig. 37

Chapter VI

Gateaux (1)

THE word gateau, translated literally, means a large tart or cake but, in this country, it has become to mean practically any cake which is decorated in a soft medium, *i.e.* not royal icing.

Generally, we expect the base to be a sponge layered with a suitable filling. The choice of decorating materials is extremely wide and an endeavour has been made in the following three chapters to employ these materials to the full, in decorative forms.

Many of the designs shown are basic, in that they can be adapted to suit almost any occasion by altering the inscription. Nevertheless, an attempt has been made to cover the usual special occasions such as 'Mother's Day', 'Easter', and 'Christmas'. The author is indebted to the Richemont Bakery School, Lucerne, for some of the ideas incorporated into these gateaux designs. The use of chocolate and jelly and of preserved fruit, for example, is typically Swiss.

'Mother's Day' Gateaux

With the exception of the last buttercream-coated gateau, all the examples shown have fondant-covered tops with sides masked in buttercream and toasted nuts.

Fig. 38 – Here we have a spray of flowers piped in chocolate with the outline of the flowers filled with red jelly. The bird carrying the spray is also piped in chocolate. Border decoration is simply a pink fondant line edged with chocolate.

Fig. 39 – This is similarly executed with a border of pink fondant outlined in chocolate.

Fig. 40 – The lily of the valley is executed in a chocolate outline of stem and leaves and the flowers are piped in white royal icing.

White royal icing and chocolate is used for the border.

Fig. 41 – Tulips are depicted on this gateau and are piped directly on to the top. These, together with the stem leaves and wavy lines, representing the ground, are all done in royal icing. The colours of the tulips are red, yellow, and mauve. The butterfly is outlined in chocolate and filled in with yellow jelly. Pink fondant, overpiped with chocolate, is again used for the border.

Fig. 42 – Use is made of crystallized rose petals to give the massed-spray effect of the vase of flowers. The vase is of pink marzipan and moulded in relief.

Fig. 43 – The flowers in this example are selected crytallized violets, the centre being composed of seven crystallized mimosa balls, and the stem and leaves cut from angelica. A small, almond-paste plaque carries the inscription.

Fig. 38
Fig. 40
Fig. 42

Fig. 39
Fig. 41
Fig. 43

63

Fig. 44 – Here we see how contrast of form can be achieved by using a bar of green-textured almond paste to break up the free-flowing line of the flower forms. The flowers are made from an orange-red 'crystallized' almond paste and have a centrepiece of a cherry. Leaves and stems are in chocolate and jelly.

Fig. 45 – An alternative way: this contrast can be effected with two bars of almond paste. Pink-coloured paste has been used here, with flowers in pale mauve and a mimosa ball for a centrepiece.

Fig. 46 – Prefabricated paste and jelly flowers make an ideal spray for the centre of this gateau. By altering the inscription on the tag, this design could be used for birthdays or almost any occasion.

Fig. 47 – The yellow primrose in this design is moulded out of modelling almond paste. With yellow flowers and green stem and leaves, this looks very attractive on a pink fondant top.

Fig. 48 – A gateau, which would be equally suitable for Valentine's Day, is illustrated here. The heart-shape is made from pink almond paste filled with chopped glacé pine-apple or pineapple conserve. Angelica is used for the stem and leaves at the top. Green is used for the base colour and it is overpiped with pink and chocolate for the border.

Fig. 49 – In this gateau, the top and sides have been coated with buttercream. Detailed instructions on how this should be done are given in Chapter IX. A special scraper has been used to achieve the ribbed impression on the sides and the bottom edge is masked in jap crumbs. The top is textured by marking with a knife fanwise from a point on the edge, and this is covered by a half-flower made from petals cut from a chocolate sheet and a cherry for its centre. The design is completed with small, piped, chocolate hearts spaced on top and around the inscription.

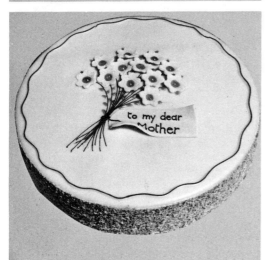

Fig. 44   Fig. 45
Fig. 46   Fig. 47
Fig. 48   Fig. 49   65

Easter and Summer Gateaux

All the gateaux in these figures have a fondant-coated top. With the exception of the three which have the sides coated in buttercream and masked in nuts, they also have chocolate-covered sides.

Fig. 50 – Here sugar-preserved pineapple has been cut into wedge shapes and inserted into a cut cherry to imitate a flower. This is assembled at intervals around the border, connected with a chocolate line, and finished off with two diamonds of angelica.

Fig. 51 – Another arrangement of the pineapple is shown here, again, using a cherry as a centrepiece with angelica diamonds.

Fig. 52 – Here cream-coloured buttercream has been used for the piped lines, which are overpiped chocolate, also the outline of the heart shapes. These are filled with pineapple conserve or crushed preserved fruit and completed with two diamonds of angelica. The centrepiece is made from prefabricated chocolate filigree petals and silver dragees for the centre. Executed on a pink-coloured base, this gateau looks most attractive.

Fig. 53 – The basket design is effected by piping a latticework of chocolate on which is laid a piece of cut pineapple confiture. Chocolate and jelly flowers complete the design.

Fig. 54 – Royal icing has been used here for the flowers, chocolate for the stem and border, and pink fondant for the bow.

Fig. 55 – The violets here are prefabricated, piped flowers and are assembled in the form of a spray as shown. The border of chocolate is embellished with silver dragees.

Fig. 56 – Again sugar-piped flowers are used with diamonds of angelica to form the floral basket. The rest of the decoration is executed in jelly and chocolate.

Fig. 57 – Another arrangement of confiture pineapple is here shown, terminating in a bulb of buttercream on which is piped a ring of red-coloured piping jelly. Diamonds of angelica are again used.

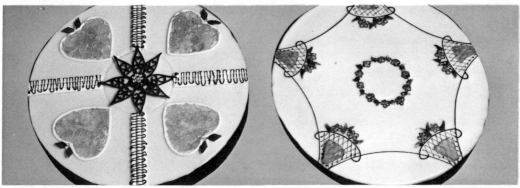

Fig. 50   Fig. 51

Fig. 52   Fig. 53

Fig. 54   Fig. 55

Fig. 56   Fig. 57   67

Chapter VII

Gateaux (2) – Stencilled Inscriptions and Gateau Tops

Stencilled Inscriptions for Gateaux

A useful size for these gateaux is approx. 3 in. by 6 in. It is suggested that the gateaux be made from two sheets of sponge baked on the normal 30 in. by 18 in. baking sheet. Using this size, such a sheet can be cut into 30 gateaux. The two sheets may be sandwiched with an appropriate filling prior to cutting.

Briefly, the technique for producing the stencilled inscriptions is as follows:

1 – Pin out thinly an appropriate paste of the chosen colour. In the illustrations only two colours were chosen—pale cream and chocolate. These two colours contrast sufficiently well with most of the coating colours, but for some an alternative pastel shade might be more appropriate.

The paste may be made of almonds or another type of nut such as coconut, or merely sugar paste. Use a special board for this purpose so that the finished paste inscriptions may be removed for setting.

2 – Place the stencil mat over the paste and trim off excess.

3 – Using an appropriate contrasting coloured royal icing, spread it through the inscriptions cut in the mat (**Fig. 58**).

4 – Carefully remove the mat to leave the inscriptions on the paste underneath (**Fig. 59**).

5 – Caramel cutters, in which the cutter blades are suitably spaced by means of spacer blocks (see later), are now used to divide the inscriptions. In **Fig. 60** the inscriptions have already been divided in one direction and are now being divided in the other at right angles.

Fig. 61 shows the finished set of inscriptions which are now allowed to harden off before being separated for use. For illustration purposes this mat contains 18 separate names, but for commercial use it is suggested that there should be a separate mat for each name containing as many inscriptions as the trade demands. The largest size mat handled by the author for this purpose covers the area of the baking tray, i.e. 30 in. by 18 in.

Fig. 58   Fig. 59

Fig. 60   Fig. 61

The Cutters

Fig. 62 illustrates five of these cutters with one half assembled showing the spacer blocks and round blades which slide over the shaft. These cutters are metric and the stencil mats have to be designed in conjunction with these sizes, i.e. in multiples of $\frac{1}{2}$ cm. The length of the inscription will also dictate the spacing of the cutter blades, and if the photographs in **Figs. 60 and 61** are closely examined it will be seen that the inscriptions on one side are longer than the other. In **Fig. 62** the two cutters on the left-hand side are assembled for cutting the gateau tops illustrated later in this chapter, whilst the two in the centre of the photograph are the ones used for cutting the stencilled inscriptions just described. For the purpose of these examples, the cutter blades were spaced to cut the paste into the following dimensions:

INSCRIPTIONS width 1 in. approx.
length $2\frac{3}{8}$ in. $+ 2\frac{5}{8}$ in. approx.
GATEAU TOPS width $2\frac{7}{8}$ in. approx.
length $5\frac{7}{8}$ in. approx.

Now that we have an inscribed plaque by which the gateau may be identified, we can give some thought to the methods by which any other decoration or finish may be applied.

The coating of the gateau may be done in a number of different ways:

1 – Enrobing with a suitable coating, i.e. fondant, fudge, chocolate, etc. This method will appeal to the large producer for its speed and efficiency with good finish.

To achieve a perfect top surface it is advisable to cover the sponge with a thin layer of almond or nut paste prior to enrobing. For the best result it is advisable to cover also with boiling apricot jam. This will not only improve the quality but prevent loss of gloss of the fondant.

2 – Coating in a butter cream or filling cream and masking the sides with nibbed or crushed nuts, corals, crumbs, etc.

The top may either be left smooth or textured with a comb scraper.

3 – Coated in a suitable cream and completely covered in nuts, corals, etc.

4 – Coating the top only in fondant or fudge, and masking the sides with cream and nuts, coral, etc.

In the following examples most of these methods of finishing are illustrated and suggestions for the introduction of the appropriate flavour given.

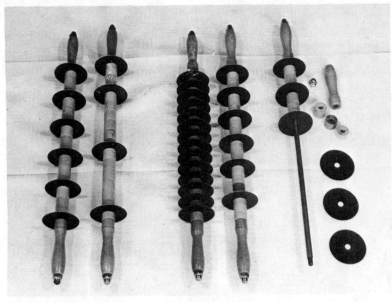

Fig. 62

Cherry – Fig. 63

Sponge base – This should contain chopped glacé cherries.

Filling – If glacé cherries are omitted from the sponge they could be incorporated here with a good butter or filling cream.

Alternatively, a good quality cherry flavour could be used.

Coating – White or pale red fondant or fudge.

Decoration – Two halves of glacé cherry with chocolate fondant piped stalks.

Fig. 63

Kirsch

A genuine kirsch gateau can only be sold at establishments making high class confectionery because of the high cost of its manufacture. For the mass market a kirsch flavoured gateau can, of course, be made using a synthetic flavour. The following ideas are for making the genuine article.

Sponge base – This may be partially soaked in a thin syrup which has been flavoured with kirsch liqueur.

Filling – A good buttercream flavoured with kirsch liqueur should be used. Tinned cherries or fresh which may have been marinaded in kirsch could also be layered in with the cream.

Coating – White fondant or fudge. This too could be flavoured with the liqueur if desired.

Decoration – Same as the Cherry gateau except for the use of the KIRSCH inscribed plaque.

71

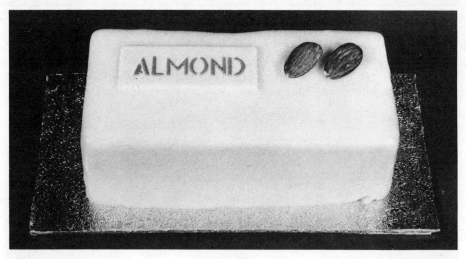

Fig. 64

Almond – Fig. 64

Sponge base – This may either be a genuine almond sponge containing ground almonds for a high class trade, or almond flavour used to produce the cheaper variety.

Filling – Here there are three alternatives:

(*a*) Buttercream or filling cream flavoured with almond.

(*b*) A spreadable paste made from almonds and sugar. Almond paste worked down with water is quite suitable for this purpose.

(*c*) The two sponge layers are sandwiched with a layer of almond paste to which it is made to adhere by the use of apricot purée.

Coating – White fondant or fudge to which almond flavour may be added if desired.

Decoration – Besides the inscription, place two unblanched almonds as shown in the photograph.

72

Walnut – Fig. 65

Sponge base – Crushed walnuts are added to the sponge prior to baking. Alternatively, for cheapness, walnut flavouring can be added.

Filling – Again there are three alternatives:

(*a*) Buttercream or filling cream flavoured with a good walnut flavouring.

(*b*) Crushed walnuts added to the cream filling. To achieve contrast in the eating quality, it is suggested that this should only be done if the sponge has added flavour. If crushed nuts have been added to the sponge base, the flavoured cream only should be used.

(*c*) A spreadable paste, made by grinding walnuts with sugar and water.

Coating – White fondant or fudge to which walnut flavour may be added if desired.

Decoration – Inscription with a selected half walnut.

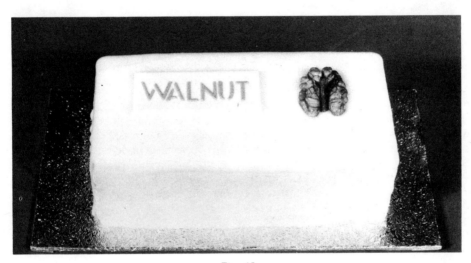

Fig. 65

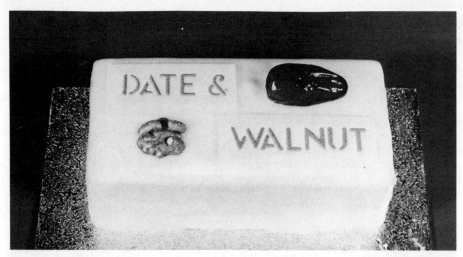

Fig. 66

Date and Walnut – Fig. 66

Sponge base – Besides crushed walnuts, diced dates are also added. The chopping of these into suitably sized pieces is made easier by using some of the flour of the mix. The walnuts can be replaced by flavouring if required, but there is no substitute flavour for the date and in any case these are comparatively cheap to use.

Filling – If no walnuts or dates have been used in the sponge, crushed walnuts and diced dates could be added to the filling cream. If they have been added, it is advisable to leave the filling cream bland.

Coating – White fondant or fudge with no flavour added.

Decoration – The figure shows an appropriate arrangement using both a whole date and half walnut.

Date

Sponge base – Diced dates added.

Filling – Here there are three alternatives:

(*a*) Add diced dates to the filling cream, if none has been incorporated into the sponge.

(*b*) Use a spreadable date paste made by mincing the dates, and boiling with water to the right consistency.

(*c*) Leave the filling bland in flavour, but only provided that diced dates are mixed into the sponge.

Coating – White fondant or fudge.

Decoration – Use of a whole date, together with the inscription.

74

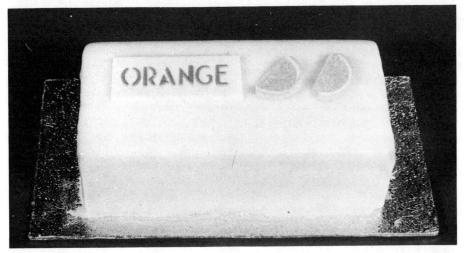

Fig. 67

Orange – Fig. 67

Sponge base – This should have a good quality flavouring which will withstand baking temperatures. The best product to use is orange paste. Besides this the sponge can be suitably coloured. A good combienne in which colour and flavour are combined may also be used.

Filling – A good butter or filling cream appropriately flavoured and coloured orange.

Acidulate slightly with citric or tartaric acid to enhance the flavour.

Coating – Pale orange coloured fondant or fudge, flavoured with orange. For supreme flavour use the powdered fondant 'Dryfon' and reconstitute with orange juice.

Decoration – Two orange jelly slices are used with the inscription as shown.

Lemon

Sponge base – Use a good flavouring, or lemon paste with yellow colour, or use a combienne.

Filling – Use a good butter or filling cream flavoured with lemon essence and acidulated with either pure lemon juice, citric or tartaric acid.

Coating – Pale lemon coloured fondant or fudge flavoured with lemon. To reconstitute 'Dryfon' use the lemon juice.

Decoration – Use two lemon jelly slices with a LEMON inscription similar to that shown in **Fig. 67**.

75

Mandarin – Fig. 68

This could be made identical in every respect to the orange gateau as shown in **Fig. 67,** except for the filling which could contain either diced or whole slices of tinned mandarin and the selected slices placed on top for the decoration with the inscription.

However a much higher quality article may be made as follows:

Sponge base – Bake without added flavour.

Then partially soak the baked sponge with a thin syrup to which has been added the liqueur Curaçao.

Filling – Flavour a good buttercream with Curaçao and after spreading on the sponge, place a layer of tinned mandarin oranges before sandwiching with the top sponge layer.

Coating – Flavour the fondant or fudge with the liqueur.

Decoration – Finish off as shown in **Fig. 68.**

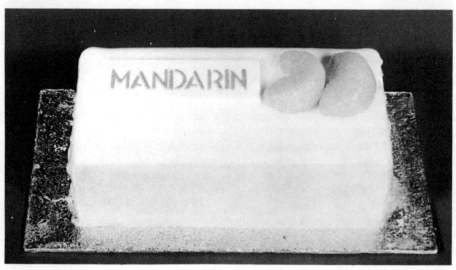

Fig. 68

76

Fig. 69

Pineapple – Fig. 69

Sponge base – This can be partially soaked with sweetened tinned pineapple juice if desired.

Filling – A number of alternatives are available:

(*a*) A good butter or filling cream to which pineapple is added in either of the following forms:

    (i) crushed
    (ii) chopped
    (iii) diced confiture (see page 9).

(*b*) Instead of using a cream, use drained and sweetened crushed pineapple as a layer.

(*c*) Make a spreadable paste by adding good quality sponge crumbs to sweetened pineapple juice and use this mixture in the same way as the filling cream.

Coating – Use a pale yellow coloured fondant or fudge. If 'Dryfon' is used, reconstitute with pineapple juice.

Decoration – Here a slice of confiture pineapple (see page 9) is used as illustrated, along with the appropriate inscription.

Other fruit gateaux which, although not illustrated here, should nevertheless be mentioned, are RASPBERRY, STRAWBERRY and BLACKCURRANT. Suggestions on their manufacture are as follows:

Sponge and Filling Cream – Use either an appropriate combienne, flavour, or the fruit extract with colour. The latter will give a superior flavour but is expensive.

When raspberries and strawberries are in season, the ripe fruit itself could be used in the filling cream for making high-class gateaux.

Coating – Fondant or fudge, flavoured with the combienne, flavouring or extract with colour. 'Dryfon' can also be used, reconstituted with the corresponding fruit concentrate. For blackcurrant – use a blackcurrant cordial or squash.

Decoration – With the inscription, use the appropriate jelly decoration for the raspberry and strawberry. The blackcurrant cannot be simulated in this way and must, therefore, be left plain with just the inscription.

Chocolate or Chocolate-Flavoured – Fig. 70

To call this a chocolate gateau the moist crumb should contain a minimum fat-free cocoa solids content of 3%.

Sponge – Some of the flour must therefore be replaced with cocoa powder to achieve the minimum fat-free cocoa solids content of 3%. The addition of chocolate colour or combienne is also desirable to achieve the right colour.

Filling – A number of alternatives may be adopted:

(*a*) Use a good butter or filling cream to which has been added:

    (i) combienne
   (ii) melted bakers' coating
  (iii) melted chocolate couverture
  (iv) grated chocolate couverture.

(*b*) Use ganache (see page 8).

Coating – Alternatives also exist for this:

(*a*) Fondant or fudge to which has been added any of the following:

    (i) combienne
  (ii) melted chocolate couverture or bakers' coating.

(*b*) Chocolate icing.
(*c*) Warmed ganache.

Other alternative combinations are:

(*a*) Use an appropriately flavoured butter or filling cream to completely mask the sponge before covering it with chocolate vermicelli.

(*b*) Leave the top with a plain cream or ganache coating, only covering the sides with vermicelli.

(*c*) Coat top only with fondant or fudge and mask sides with cream and vermicelli.

Decoration – Here only the inscription is employed but, if desired, chocolate off pieces could be used. However, care should be taken not to display such decorated gateaux or the one covered with vermicelli in strong sunlight, otherwise the chocolate will melt and spoil the appearance.

Fig. 70

78

Fig. 71

Hazelnut – Fig. 71

Sponge – Ground hazelnuts are added to the sponge to impart flavour, or alternatively a good flavouring is used.

Filling – Here there are alternatives:

(*a*) Use a good butter or filling cream to which the following is added,

    (i) a good flavouring
    (ii) crushed hazelnuts (these could be roasted first). To provide contrast, this should *not* be used if crushed nuts have been used in the sponge.

(*b*) A hazelnut paste used to sandwich the two sponge layers.

Coating – A good hazelnut flavoured butter or filling cream is more appropriate to use than fondant or fudge. In the illustration the top has been left plain but the sides are masked in the cream and crushed and dressed roasted hazelnuts. The cream used has been delicately tinted to the same colour as the roasted nut.

Decoration – Two roasted whole hazelnuts with the inscription complete the gateau.

Praline – Fig. 72

Sponge – Since praline is made from a mixture of nuts, either almond or hazelnut essence could be used in the sponge. However, the praline flavour in the filling cream and coating is sufficiently strong to dispense with flavouring in the sponge. For high-class gateaux, the sponge could be partially soaked with a rum-flavoured syrup, preferably using real rum spirit.

Filling – Use a good butter or filling cream to which is added sufficient of the praline paste to bring it to the desired flavour. If the praline is hard, plasticise it with heat before incorporating it into the cream.

Coating – The top and sides are first coated in the praline cream. The top is then textured, using a serrated scraper and the sides are masked in crushed or nibbed roasted nuts.

Decoration – The textured top provides the main decoration here and it only needs the inscription to be placed on.

Nougat

This is similar to the previous gateau since the praline is merely nougat milled to a paste.

Instead of using praline in the filling cream, crushed nougat may be added. The cream used for coating could be flavoured with praline, but the sides of the gateau are masked in nougat nibs dressed out of crushed nougat.

Fig. 72

Fig. 73

Coffee – **Fig. 73**

Sponge Filling and Coating – This may be suitably flavoured with a combienne, extract, flavour or powdered instant coffee. The sponge could be partially soaked in a rum syrup, if desired, for a high-class product.

Decoration – Besides the use of the inscription, two sweet coffee beans may be used and placed alongside in the same way as the nuts in the hazelnut gateau in **Fig. 71**.

Christmas – Fig. 74

Gateaux for Christmas may be made in a variety of flavours and finished off in a great number of different ways.

Illustrated here is only one example of the way the stencilled inscription can be used with a motif. The artificial holly leaf could be replaced by a very large variety of motifs, such as:

(*a*) Edible stencilled run-out holly leaves with piped red berries (see page 25).

(*b*) Artificial or stencilled robins.

(*c*) Miniature holly trees.

(*d*) Miniature snowballs. These can easily be made by covering glacé cherries in white sugar paste and icing sugar (see page 17).

(*e*) Miniature logs. These are made by covering a rope of almond paste in chocolate, marked with a fork prior to setting and then cut up into short lengths. Before placing them on to the gateau, dust with icing sugar for a snow effect (see page 16).

Fig. 74

Fig. 75

Easter – Fig. 75

Other festive occasions such as Easter and Mother's Day may also be used to induce sales of gateaux. Again, with the inscriptions, several motifs could be used for Easter such as chicks, rabbits, the Cross and eggs. In the illustration a little nest containing a sugar egg has been used. The nests are made from chocolate coloured royal icing piped into a circle with a star tube and dressed with chocolate vermicelli. Similar ones can be made with chocolate mixed with fine desiccated coconut. **Fig. 76** shows these being piped. Alternatively, the nest can be dispensed with, just leaving the sugar egg.

Fig. 76

Stencilled Gateau Tops

Instead of merely producing plaques with a stencilled inscription, by the same technique we can produce complete tops which not only carry the inscription but an appropriate design as well. For mass production mats should be the size of the baking sheet 30 in. by 18 in., carrying 30 impressions of the same gateau design, but for the purpose of these illustrations the mats were made smaller and carried a limited number of different designs.

The technique is described as follows:

1 – A suitably coloured and flavoured nut or sugar paste is pinned out thinly to cover an area slightly larger than the size of the stencil mat.

2 – The surface of the paste is brushed free of any dusting sugar and the mat laid on.

3 – We can now either use a contrasting colour for the impressions to produce a silhouette, i.e. chocolate on cream or white on chocolate, or we can colour various sections of the mat with the appropriate coloured royal icing. In **Fig. 77** various coloured icings are being pressed through the impressions of the stencil. For example, green is being used for leaves, orange for the flowers and chocolate for the inscription.

4 – Next, the stencil mat is carefully lifted away as shown in **Fig. 78**.

5 – The caramel cutters are now used to divide the tops, first in one direction and then at right-angles as illustrated in **Fig. 79**. To ensure that the tops are cut with some measure of precision it is necessary to trim the paste to the stencil mat, so that there is a straight edge with which to guide the cutter. The blades of the cutter were spaced to give a size slightly under the 3 in. and 6 in. chosen for the gateau, to allow for the wastage when cutting the sponge.

Fig. 80 illustrates the six stencilled gateau tops suitable for Mother's Day, Easter and Christmas. These will now be described in greater detail.

It only requires 24 hours for these tops to set sufficiently to be handled by being placed on to the gateaux, which have been previously coated with butter or filling cream and the sides masked with nuts, corals, crumbs, etc.

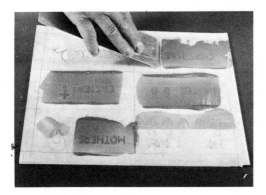

Fig. 77

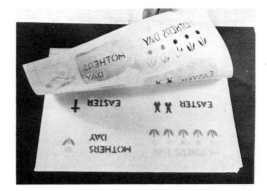

Fig. 78

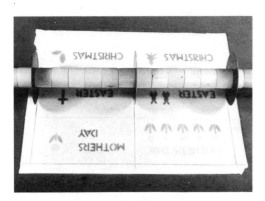

Fig. 79

Fig. 80

Mother's Day Gateaux – Figs. 81 and 82

The stencil in **Fig. 81** was executed using two different coloured icings – red for the inscription and flowers and green for the stem and leaves. The flower centres were completed by piping on a bulb of yellow icing.

To execute the design in **Fig. 82**, red and green has again been used for the flower and leaves, but chocolate for the inscription. The flower pot is cut out of chocolate almond paste, previously textured with a ribbed roller. These are quickly cut from strips alternating the angle of each cut, so that wastage only occurs at each end.

Fig. 81

Fig. 82

Christmas Gateau – Figs. 83 and 84

The tree in **Fig. 83** is stencilled in green, whilst the inscription is in chocolate. Again the cut-out flower pot is applied after the stencil has set.

The same coloured icings are used to execute the design in **Fig. 84**, except that the berries are piped in with vivid red, slightly softened, icing.

Fig. 83

Fig. 84

Easter – Figs. 85 and 86

The motif used in **Fig. 85** is the rabbit whilst that in **Fig. 86** is the Cross. This latter motif would find favour with religious customers who may take the true reason for an Easter celebration seriously.

Other motifs which could be incorporated into designs suitable for Christmas and Easter have already been suggested in describing **Figs. 74 and 75**.

Fig. 85

Fig. 86

88

Stencilled Tops for Fruit Gateaux

In the illustrations the stencilling is done in chocolate icing on a cream coloured paste, but using the technique already described it is perfectly feasible, although slower, to stencil the fruit shape in the appropriate colour.

Cherry and Kirsch – Figs. 87 and 88

The cherries in these designs may be stencilled in red, using chocolate for the stems.

Suggestions for flavouring and the make-up of the gateau has already been made (see previous Cherry and Kirsch Gateau – **Fig. 63**, page 71).

Fig. 87

Fig. 88

90

Orange – Fig. 89

The fruit in this example may be stencilled in orange colour (For suggestions on flavour see previous Orange Gateau – **Fig. 67.**)

Fig. 89

Lemon – Fig. 90

Here the fruit is stencilled in yellow icing. (For suggestions on flavour see previous Lemon Gateau, page 75.)

Fig. 90

Apple – Fig. 91

Green may be used here for the apple impression.

Sponge – An apple essence may be added to the sponge.

Filling – Instead of a cream use a thick well-boiled apple purée. Use whilst still warm so that it sets firm.

Coating – Apple purée, jam or conserve masked with toasted coconut.

Fig. 91

Pear – Fig. 92

Green may be again used for the impression of the pear.

Sponge – Add a pear essence to the sponge.

Filling – Employ a purée from pears and add pectin to make it thick and use whilst still warm.

Coating – Pear purée, jam or conserve masked with toasted coconut.

Fig. 92

92

Banana – Fig. 93

The banana impression is executed in yellow sugar.

Since this fruit, once cut, readily browns and spoils its appearance, it cannot be used in its natural state to impart flavour. Fortunately, there are a number of excellent banana flavours on the market and we have to rely upon these to impart the banana flavour into the sponge and cream filling. The sides may be masked with banana flavoured butter or filling cream and either roasted nuts, corals, or crumbs.

Fig. 93

Apricot – Fig. 94

The fruit may be stencilled in apricot coloured icing and the leaves and stem in green.

Sponge – This may be flavoured with a good essence and coloured yellow.

Filling – There are two alternatives:

(*a*) Use tinned apricots with a good quality cream.

(*b*) Use a good apricot conserve.

Coating – A good butter or filling cream using toasted nibbed almonds for masking the sides.

For a high-class gateau, apricot brandy could be used with stock syrup to partially soak the sponge and flavour the cream for filling and coating.

Fig. 94

Chocolate Flavoured – Fig. 95

The stencilled top here is merely a chocolate coloured almond paste with the inscription stencilled in white icing.

For suggestions regarding flavour and make-up see the previous Chocolate Gateau – **Fig. 70**, page 78.

Fig. 95

Round Stencilled Gateaux Tops

The use of almond or sugar-paste tops, suitably decorated with a stencilled design, has obvious advantages. These may be made and stored by the hundred, to be used when the occasion demands, such as for festive occasions like Easter and Christmas. The technique for mass producing these gateaux tops is as follows:

The paste is pinned out to the desired thickness, the stencil mat laid on, and the design stencilled through. A large circular cutter is used to cut out the disc of paste which may be transferred directly to the gateau or on to a board for storing.

Although the production of only one top is explained, obviously this should be done in units of, say, 12, the paste being pinned out to cover a large area and the stencil mat used 12 times before being removed for cleaning ready for the next unit operation.

If the tops are required to be stored, the cutting-out could take place on the board on which they are to be stored. This eliminates double handling. Twelve such tops of 7 in. diameter can be cut out and stored on a board 2 ft. by 2 ft. 6 in. A rack with an inch clearance between boards could, therefore, store a considerable number of these stencilled tops to be used as the occasion demands.

If desired, the edge of these tops can be crimped either by hand or with the nippers.

Easter Gateaux

Figs. 96 and 97 – The first two illustrated are stencilled in chocolate-coloured royal icing and, when set, the catkin flower is piped in white. Whilst still wet, these flowers are dredged with castor sugar.

Fig. 98 – This gateau is similarly executed but this time the rabbit is also dredged with sugar. Green is stencilled through for the grass.

Fig. 99 – Use of differently coloured icings is demonstrated in this figure. The colours used are, chocolate for inscription, butterfly body, and stems; red for the flower petals; and yellow for the flower centres and butterfly wings.

Notice that, in all these illustrations, chocolate buttercream has been used with coralettes for the sides. This provides a pleasing contrast between the top and the sides of the gateau.

In all these gateaux the tops have hand-crimped edges.

Christmas Gateaux

Fig. 100 – Chocolate-coloured icing has been used to stencil the Father Christmas in this first gateau top. The inscription is piped in freehand afterwards. Metal nippers are used to effect the edge shown on this top.

Fig. 101 – Contrast is created in this figure by stencilling white icing on to a chocolate almond-paste top. The holly leaves are tinted with green colour and the flame with orange. Berries of red icing are piped on the holly leaves when dry. Because of the dark coloured top, a light coloured side is executed, by using lightly toasted coconut on buttercream.

Fig. 102 – These stencils can, of course, be used directly on to the cake and the Christmas tree design has been used in this way. Again, different coloured icings are used; green for the branches; chocolate for the tub and inscription; white for the candles; and yellow for the star and flames.

Fruit Gateaux

Fig. 103 – The wild strawberry design is executed in the same way using red, green, white, yellow, and chocolate-coloured icings. No attempt is made here to pattern the edge which is left just as it emerged from the cutter.

Fig. 98
Fig. 100
Fig. 102

Fig. 99
Fig. 101
Fig. 103

Chapter VIII

Gateaux (3) – Christmas

IT is a fact that the Christmas gateau is becoming very much more popular with the public, many of whom seem to prefer this type of celebration cake to the traditional cake decorated in royal icing. There is tremendous scope for the confectioner in decorating this type of cake, as the illustrations in this chapter portray. The growing importance of this type of gateau merits devoting a chapter to its exposition.

Exceptions were made with the three described and illustrated in the previous chapter, because of the special technique involved. It was felt best that these should be illustrated and explained with the other, prefabricated, stencilled, gateau tops.

Most of the cakes illustrated are decorated with prefabricated pieces.

The first two have fondant-covered tops and sides coated in chocolate couverture.

Fig. 104 – The snowman is executed in fondant with an inscription in chocolate and this gateau has a border of cherry snowballs.

Fig. 105 – This gateau is decorated in chocolate with run-out prefabricated Christmas tree branches, linework and inscription. A white-fondant-piped candle with a strip-almond flame completes the design.

Figs. 106 and 107 – The next two gateaux employ an almond-paste chalet as the focal point of the design. The chalet is made from six parts, two gable ends, two sides, and halves of the roof. These latter parts are made from almond-paste which has been textured with the nib roller, to imitate the tiles.

The gable ends can be quickly made by cutting a hexagon-shaped piece in half. These six parts are cemented together with royal icing and, when set, the chalet is dusted with icing sugar and mounted on to the irregularly shaped base, also cut out of paste.

The Christmas trees are made by cementing two cut-out paste shapes together with royal icing and supporting them in a bulb of royal icing until set, so that they will stand vertically on the gateau. Both employ a paste plaque to carry the inscription, one of which is stencilled whilst the other is piped freehand. The holly leaves used are also prefabricated.

Although the design is basically the same for each gateau there are two very different effects. In **Fig. 106** the whole top and sides are covered with a fudge icing with an edge of roasted nuts. The top of this icing is roughened to give a snow impression. The other gateau, in **Fig. 107**, is covered in chocolate butter-cream and vermicelli, with a dusting of icing sugar on top.

Figs. 106 and 107

Figs. 108 and 109 – The design of these two gateaux is basically the same. Coated in chocolate buttercream they are masked in chocolate vermicelli and vertical lines piped on in white cream. The gateaux are now finished off with prefabricated decoration. **Fig. 108** has freehand-cut almond paste holly leaves with small alpine cherries for the berries, whilst **Fig. 109** has candle and holly-leaf cut-outs.

Fig. 108

104

Fig. 109

Fig. 110

Fig. 110 – This next gateau, in **Fig. 110**, is first coated with chocolate buttercream and then completely masked in milk chocolate vermicelli. A strip of almond-paste, on which the inscription has been stencilled, is laid on to overlap the sides. A group of stencilled holly-leaves and three Christmas trees completes this simple design.

106

Fig. 111

Fig. 111 – In this figure warm fudge icing is used to cover the base and the bottom edge is masked in chocolate corals before it has set. A disc of textured almond paste is then placed on the top and on this is arranged the inscription and the other motifs of holly and Christmas trees.

Fig. 112

Figs. 112 and 113 – Yuletide logs are a popular decoration to use for decorating Christmas gateaux and, in the first two illustrations, we see how these may be used. In the first example, **Fig. 112**, they are placed on to a fondant-topped gateau, the sides of which are masked completely in chocolate. Two plastic robins are also used.

The second example, **Fig. 113**, shows an arrangement with cherry snowballs and run-out chocolate Christmas tree branches. This gateau, and most of the others illustrated in this and the following illustrations, are completely covered in a warm fudge icing. When set, the bottom edge is covered with chocolate.

Fig. 113

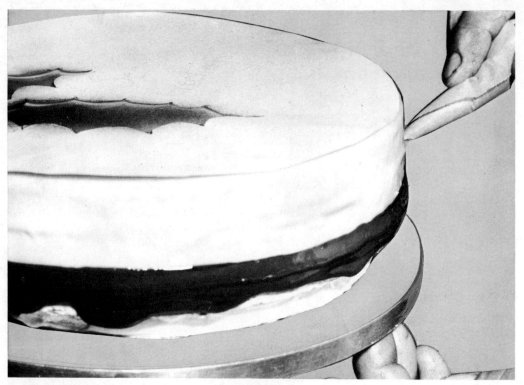

Fig. 114

Fig. 114 – The chocolate is applied from a bag held close to the bottom edge whilst the gateau is spun on the turntable, so that the molten chocolate runs down. Before it is completely set, it is neatly trimmed off.

Fig. 115–In the next gateau, the air-brush has been used to stencil green colour in the form of a holly spray on to the fudge top. This is shown in the previous illustration. The holly shape is outlined in white buttercream and on it is laid a prefabricated chocolate holly form. The berries in this case are made from small alpine cherries, but bulbs of red piping jelly would look just as effective. Almond paste plaques carry the inscription in chocolate.

Fig. 115

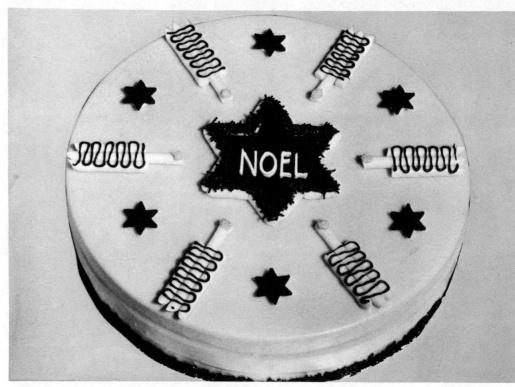

Fig. 116

Fig. 116 – This design also shows the use of the stencil. This time a star shape has been stencilled in chocolate colour, outlined in white buttercream and covered with a pre-fabricated vermicelli-coated chocolate piece. The chocolate background shows up the inscription which is executed in white royal icing. Other stars of cut-out chocolate are used, and sugar balls introduce some much-needed colour into the design. Parallel piped buttercream lines overpiped with chocolate bring in contrast of form.

Fig. 117 – Chocolate vermicelli dressed Christmas trees are the main theme of this design. These are lightly dusted with icing sugar before being laid on to the gateau. The almond paste plaque in the centre is partly textured with white royal icing to imitate snow. Nibbed almonds are lightly sprinkled over the top to break up the severe lines of the design and provide some contrast. The bottom edge in this case is chocolate vermicelli which is applied to the fudge coating before it sets.

Fig. 117

Fig. 118

Fig. 118 – Chocolate is used exclusively for the decoration of this gateau with the motifs prefabricated and placed on. Toasted almonds are used for the bottom edge.

Fig. 119

Figs. 119 and 120 – Star shapes run out in chocolate directly on to the fudge top form the decoration of the following two gateaux. The first of these, **Fig. 119**, also has a top textured with a knife so that the lines radiate from a point on the edge. Surrounding this is placed a plaque, to carry the inscription and to enclose three prefabricated Christmas tree branches. The gateau, in **Fig. 120**, is more simply decorated.

Fig. 120

Fig. 121 – A stencil has been used in this gateau to outline the tree shape in green. This is outlined in white buttercream.

Circles are cut into the top, which, in this instance, is fudge but which could be almond paste and into these is run orange piping jelly. White lines of buttercream or royal icing is piped from the tree outline to meet these jelly circles and a flame of yellow is piped over. An almond paste tub and chocolate inscription completes the decoration.

Fig. 121

117

Chapter IX

Torten

ON the Continent almost any gateau, large tart, or flan, is called a torten, but in this country we have come to accept a torten as a large decorated gateau which can be divided in such a way that wedge slices can be cut out without breaking up the overall design. Thus, each individual slice is designed and contributes to the design of the torten as a whole.

The main idea of this form of design is so that individual slices may be sold over the counter, or served at coffee, or as an after-dinner sweet. The interior, therefore, is also worthy of consideration, not only from the eating point of view, which is so important, but also from that of the design.

Our continental colleagues go to a considerable amount of trouble to get design and contrast of colour into the interior of their torten, by using different types of sponge and sponge and jap base, and combining the flavour with these intelligently. Cream is invariably used as the carrier for the flavour and, in the first chapter, not only is a recipe given but also details of how such a cream may be flavoured. It is essential that a good cream, made preferably of butter, be used to carry the flavour, since this does not only impart a delicious flavour of its own but is also an ideal spreading medium. The following section of a torten is recommended. It should never be more than 2 in. in height and may vary in diameter from 8 in. to 11 in.

Some details of the technique of assembling and coating these torten in buttercream are given here.

A baked jap disc is placed on to a circular torten card, and given a coating of raspberry or other jam. The torten card is made in waxed cardboard and should have a size about

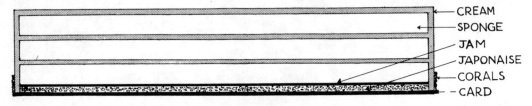

SECTION OF TORTE

Fig. 122

cream fillings and introducing fruits and nuts, etc. It should be remembered that contrast of flavour and texture is just as important, with the interior of the torten, as contrast of colour and texture is in the decoration. Thus, a layer of sweet jam, in an otherwise bland cream, gives the palate a delightful sensation. Also the layer of a crisp crunchy material, such as jap, goes well with a soft sponge.

To make good torten it is worth while using good materials, making up a special

$\frac{1}{2}$ in. larger in diameter than the jap and sponge bases: cake strawboards may also be used for this purpose, although they will prove more expensive.

The sponge is now assembled on to the japonaise in three layers, with a liberal layer of flavoured cream between. This sponge may be baked either in one piece and cut into three afterwards, or it might be more convenient to bake three very thin sponges.

After assembly, a cakeboard is placed on

top and pressure applied to ensure a perfectly flat top surface. The base is now completely masked in buttercream. Using a celluloid scraper, plain or serrated, the side is coated by holding the scraper against the edge of the torten card or strawboard, so that an even layer of cream is perfectly applied. The card acts as a template in this connection.

Using a long wide-bladed knife, the top is levelled by sweeping the knife from the outside edge to the centre from about four positions of the turntable. After practice, a level and smooth top can be accomplished: a clear unbroken edge is the aim.

An alternative method is to coat the top, first by sweeping over with a knife or straight-edge and cutting off with a palette knife as the sides are coated. This, however, does not give such a smooth and perfectly round side coat and, in the author's opinion, does not give such a good finish to the torten.

By means of the card under the torten, the latter may be picked up without sagging and any edge or side decoration applied. The card is intended to stay with the torten and is, therefore, sold with the cake to the customer, unless the torten is sold by the slice. In this event, special card containers are often supplied, to ensure that the customer gets the torten home without damage.

Christmas Torten

Fig. 123 – The first design illustrated is typical of a continental torten. Lines of buttercream are piped radially from the centre on divisions which have been previously marked with a plastic torten marker. The centre of these torten is usually textured with coloured nibs, vermicelli, jap crumbs, or a mixture of cocoa-powder and icing sugar, and is applied through a circular cutter around which has been made a mask of stiff paper as seen in

Fig. 124 – Surrounding this textured circle and resting on the ends of the piped lines is a circle of almond paste, which carries the inscription. This circle of paste is prefabricated and allowed to set before being used in this way, otherwise it would tend to sag. Chocolate is used to prefabricate the motifs used and a light sprinkling of toasted nib almonds helps to break up the design.

Fig. 125 – This torten is completely masked in chocolate buttercream and vermicelli. The centre is dusted with icing sugar and white cream lines are piped from this centre to the edge. Laid alternately on to this are almond paste cut-outs of holly and Christmas trees.

Fig. 126 – Another Christmas torten. The top is covered with a chocolate icing and sides masked in cream and jap crumbs. Cherry snowballs are used for the decoration and these are placed on to white buttercream lines piped from the centre. A paste-moulded snowman makes an apt centrepiece.

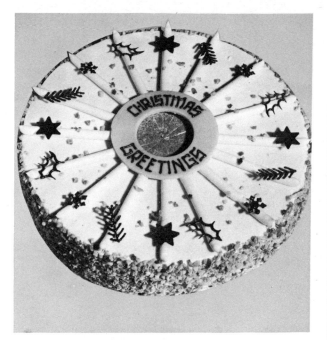

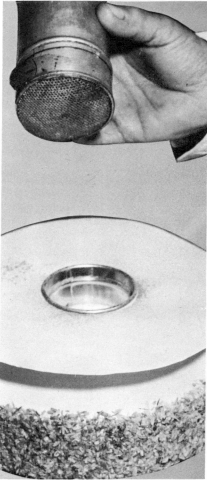

Fig. 123
Fig. 125

Fig. 124
Fig. 126

121

Figs. 127 and 128 – The next two 'Mother's Day' torten designs have attractive side finishes. They are both completely covered in buttercream and decorated in chocolate cut-out and run-out shapes. Toasted coconut is used for the masking of the bottom edge. The inscription in **Fig. 127** is carried on an almond paste ring, whilst in **Fig. 128** a chocolate disc is used.

Easter Torten

The next four examples would be suitable for Easter torten. For the first two pieces of pineapple confiture with chocolate and jelly on a fondant top are used. **Fig. 129** has sides of cream and nuts whilst the torten in **Fig. 130** is completely covered in chocolate.

Figs. 131 and 132 – The next two torten utilize filigree chocolate half-flower and petal shapes with white buttercream, angelica, and silver dragees on a chocolate-cream-coated top. Sides are masked in toasted almonds and coconut.

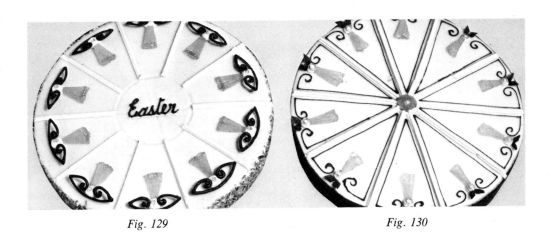

Fig. 129   Fig. 130

Fig. 131   Fig. 132

Stencilled Almond Paste Tops

Fig. 133 – Stencils and the plastic torten marker provide us with the means to produce prefabricated torten tops for any occasion. This technique is illustrated in six stages in this figure and is hereby explained.

The paste is first pinned out to about $\frac{1}{16}$ in. thick, and on it the stencil is placed. Coloured royal icing is spread through the appropriate parts of the stencil. In the photograph, the green stem and leaves have already been stencilled in green sugar and white is being spread through the flower impressions. The stencil mat is now carefully removed.

Next a plastic torten divider is positioned accurately over the stencilled design and pressed firmly down so that it cuts the almond paste into the divisions – 16 in this particular case. The paste is trimmed to size either by cutting with a large circular cutter, after removing the stencil mat, or by running the knife blade around the outside of the plastic divider. An ordinary round cutter is used to cut out the centrepiece. The torten marker is pressed on to the cream top of the torten and, into the divisions thus marked, are placed the individual sections.

Fig. 133

Examples of Stencilled Designs

Fig. 134 – The red flower forms illustrated in this first example have centres of piped yellow sugar, but these could be mimosa balls, as shown in the last photograph. Stencilled sugar-dredged petal run-outs are used for the centrepiece. The sides of this and the following torten are of cream and nuts.

Fig. 135 – Apricot yellow and green are used to stencil the apricots or peaches in this torten. The fruit forms are sprayed pink with the air-brush and a stencilled run-out butterfly is used as a centrepiece.

Top, Fig. 134   Bottom, Fig. 135   129

Fig. 136 – In this next torten, the Easter motifs previously used for the fancies are employed. These motifs are stencilled in pairs but, obviously, by using them individually, eight differently designed tops may be made. Chocolate cream and vermicelli have been used for the side of this and the successive examples.

Fig. 137 – This torten embodies the designs used in the Christmas fancies, in the same way, with a centrepiece of white royal icing dredged with castor sugar.

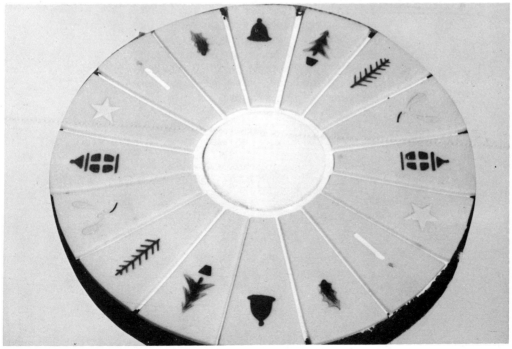

Fig. 138 – The fruit shapes are used in this next design. It would be best, of course, to use only one or two of these fruit shapes together but this shows an example of the use of nine different shapes.

Fig. 139 – Heather and thistle are depicted in this last example. Stencilled in green, they are touched up afterwards with mauve from a brush. The centrepiece tartan is made out of piping jellies. First, the almond paste circle is cut out and coated with a pale yellow jelly. Thin lines of red and green are then piped over in the pattern of a tartan.

This torten is more elaborate than the others but looks most attractive.

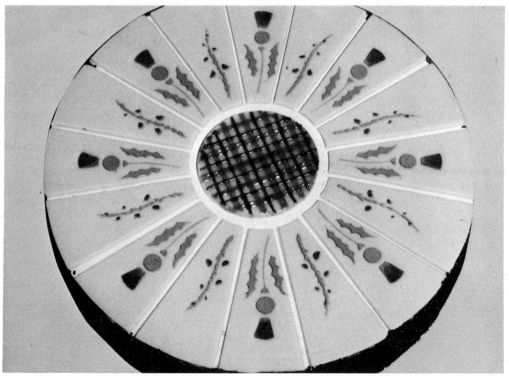

Chapter X

Battenburgs, Layer Cakes, Slices and Marbling

Battenburgs

Before giving examples of the decoration of battenburg cakes, some notes on their preparation from the genoese base might be helpful. Whilst it is possible to make battenburg patterns of 9 and 16 squares, with variations, these are in the exhibition field and so are not dealt with in this chapter. The preparation of the four-square variety consisting of pink and white genoese is explained.

A sheet of white genoese and one of pink is trimmed to remove the crust and caused to adhere to each other by the use of boiled apricot puree. The jam should be of good quality and applied liberally. It should contribute to the flavour of this type of cake, as well as being the medium for holding the various portions of coloured genoese together. The double sheet of genoese is cut into strips about 1 in. wide and each strip laid over on its side. After the whole sheet has been cut and so treated, boiled apricot puree is again applied liberally. The strips are reversed then cemented together white to pink to form the end checkerboard pattern.

Almond paste is pinned out to about $\frac{1}{8}$ in., trimmed to the same width as the genoese strip and liberally covered with puree. The composite strip of genoese is laid on to the almond paste at one end and turned over four times, picking up the almond paste and so being completely enveloped. A knife is used to cut the strip free from the rest of the almond paste and a second strip is covered in paste in the same way. Several pieces may be so wrapped before the almond paste is used

up and another piece pinned out. Any paste remaining, which is insufficient to cover at least one side of a strip, may be worked into the next almond paste piece or kept for use in preparing prefabricated decorations.

The strips of battenburg may now receive an edge decoration, by crimping either by hand or with the use of special nippers. Whilst the use of the latter can impart some very attractive designs, the decoration is very much more slowly formed, as it has to be effected a nip at a time.

Hand-crimping is more commercial, since the decoration is applied on the two edges simultaneously. Also with hand-crimping the fingers are much kinder to the paste, with the result that even a very thin paste coating may be crimped which would be impossible with the nippers.

Two other treatments which may be given are as follows:

1 – The paste may be textured with a special patterned roller prior to wrapping around the genoese strip.

2 – The top of the battenburg may be marked with a knife or special shaped cutter before crimping the edges.

This is the most decoration that is usually applied to commercial battenburg. However, for very little additional effort, attractive decoration may be added which will certainly increase the popularity and sale of these cakes. Six examples of decorated battenburgs are now given.

Fig. 140 – This battenburg is hand-crimped. the top decoration consists of half cherries and angelica diamonds arranged alternately.

Fig. 141 – Another hand-crimped variety but with a more elaborate decoration. A wavy line of chocolate or chocolate-coloured fondant is first piped with branches to join the almonds. These are split almonds, the lower half of which is dipped into chocolate and allowed to set before being placed on to the battenburg. Diamonds of angelica are again used to give the floral effect.

Fig. 142 – This is a similar design to the previous figure except that almond-paste prefabricated flowers are used. The edges are crimped with nippers.

Fig. 143 – The same edge decoration is applied to this battenburg as that in **Fig. 142**, but the top decoration consists of prefabricated, chocolate-filigree, flower shapes, with mimosa ball centres.

Fig. 144 – This and the following example have a differently patterned edge applied with nippers. In this figure we see the use of crystallized almond paste cut-outs to form the top decoration.

Fig. 145 – Special occasions, such as Easter and Christmas, may be depicted by the use of a suitable motif. Here we see how stencilled sugar holly leaves may be used in this way.

Layer-Cakes

The make-up of these cakes is different in that, instead of the checkerboard pattern, the interior is built up of layers. These may be of the same genoese or of layers differently flavoured and/or coloured. The layers may be sandwiched together with a variety of mediums, such as jam or puree; buttercream; curd; jelly; ganache; or truffle cream. Whatever the medium, it should be applied liberally so as to contribute to the overall flavour of the cake.

The layered strip may now be covered in a wide range of different materials such as almond paste; sugar paste; fondant; buttercream; fudge; chocolate; or a combination of two or more of these.

Fig. 146 – The two cakes illustrated here are first completely masked in buttercream and the sides dressed in roasted flake almonds. That on the left has heart shapes outlined in cream and filled with an orange jelly. Diamonds of angelica are also added.

The example on the right has a buttercream piped border and centre scroll design. The overpiping of this, and the wavy edging, is done in orange jelly and red sugar balls are added.

Fig. 147 – Both these cakes have buttercream-coated sides masked in toasted coconut and a thin almond paste top. The cake on the left has a border piped in buttercream to retain a marbled fondant top. The example, on the right of the figure, is decorated by first piping spirals of chocolate across the centre, laying on two ropes of coloured almond paste and filling in with an apricot jelly. This type of design is suitable for the layer cake to be divided into slices.

Fig. 140
Fig. 142
Fig. 144

Fig. 141
Fig. 143
Fig. 145

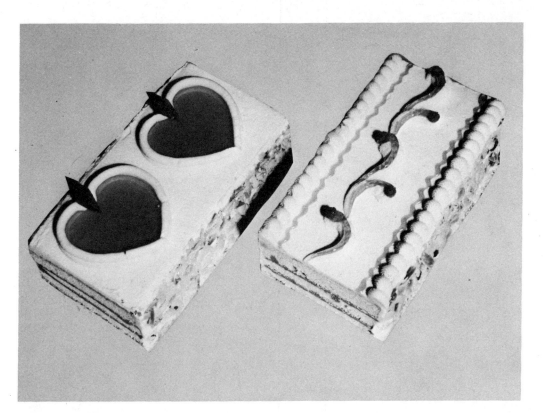

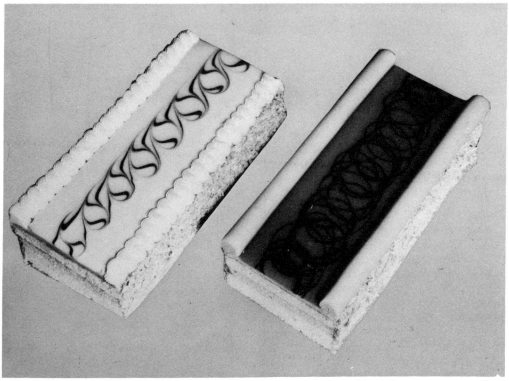

Fig. 148

Fig. 148 – Fondant is used for coating these two cakes. That on the left has a decoration marbled into it as described and explained in Chapter XII. A mimosa ball forms the centre of the flower. Glacé pineapple pieces and cherries form the floral design of the other cake. The line is piped in chocolate or chocolate-coloured fondant and angelica is used to give the leaf impression.

Slices

If made small enough, layer cakes may be cut into slices and, if the decoration is suitably arranged, each slice may be decorated to produce very attractive fancies. These are made much more speedily than the totally enrobed variety and are, therefore, a very good commercial proposition.

Besides the rectangular section, strips may be triangular or semi-circular in shape.

Buttercream-covered Slices

Coating a strip of genoese with buttercream involves much time and rarely can one achieve a good edge. The use of a special scraper which can impart a smooth surface to three sides at once is shown in **Fig. 150 (a)**. This may be easily made by bending a sheet of aluminium to the size and shape required. The buttercream is spread liberally and then the scraper drawn down the length of the slice in one movement, as shown. The size of the scraper should be $\frac{1}{4}$ in. larger in the width and $\frac{1}{8}$ in. greater in depth than the slice, so that $\frac{1}{8}$ in. layer of cream is left all over. Other shapes may also be made from this pliable metal as shown in **Fig. 149**:

Examples of slices in which these other scrapers are used will be shown later.

Having coated the slice, the next operation

is to divide it quickly and accurately into the number of slices required. This is done by rolling over it the caramel cutter as illustrated in **Fig. 150 (b)**, the blades having been set to the correct width by inserting the appropriate size blocks. The strip is now ready to receive the decoration which is placed in the centre of each slice. For commercial use a fudge cream is recommended since this skins and enables it to be handled more easily.

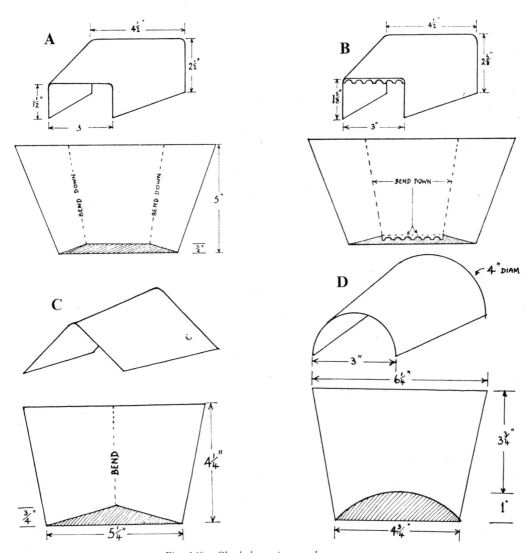

Fig. 149 – *Shaded portions to be cut away*

Although the sizes given may be altered to suit individual requirements there are two important considerations:

(1) The leading edge of the scraper should be larger than the cake by only about ⅛ in. all round so that only approximately ⅛ in. of cream is spread on.

(2) The walls of the scraper must taper away from the leading edge so that it spreads the cream at about an angle of 15°–20°.

139

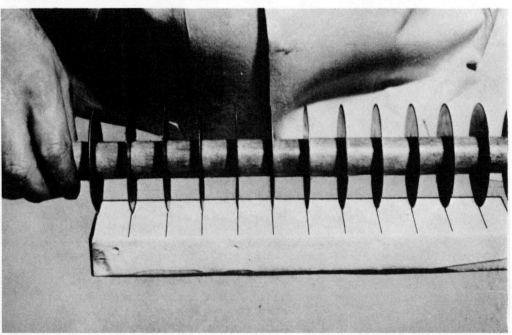

140   *Top, Fig. 150(a)*   *Bottom, Fig. 150(b)*

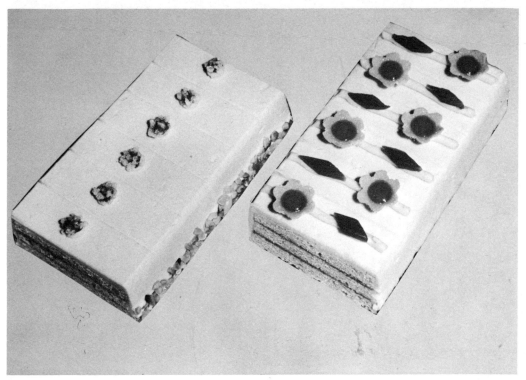

Fig. 151

Fig. 151 – Both examples here are covered with buttercream, using the rectangular scraper. That on the left has a bottom edging of roasted nib almonds, and red sugar bulbs dressed in nib sugar form the centre decoration. The right-hand example is more elaborate with jelly-filled almond-paste flowers and angelica diamonds arranged alternately on a line of buttercream. Chocolate vermicelli is used for a bottom edge.

Fig. 152

Fig. 152 – Here the decoration is applied to a boat scroll, piped in buttercream. The left-hand example shows the use of crystallized almond paste cut-outs, whilst the other cake is decorated with glacé cherries and pine-apple jelly slices cut and arranged as a flower, with a diamond of angelica for the leaf impression.

Fig. 153 – The use of scraper pattern B is illustrated in this photograph. Both have a bottom edging of chocolate vermicelli mixed with green nibbed almonds. For the cake on the left the decoration consists of diamonds of angelica and red sugar balls. Chocolate is spun over to form the main decoration of the other cake which also has a few pieces of broken violet petal placed in the centre.

Fig. 154 – Pattern C scraper is used here to cover the base with buttercream. The use of the stencilled sugar chick and holly motifs illustrate how slices may be decorated to suit festive occasions, such as Easter and Christmas.

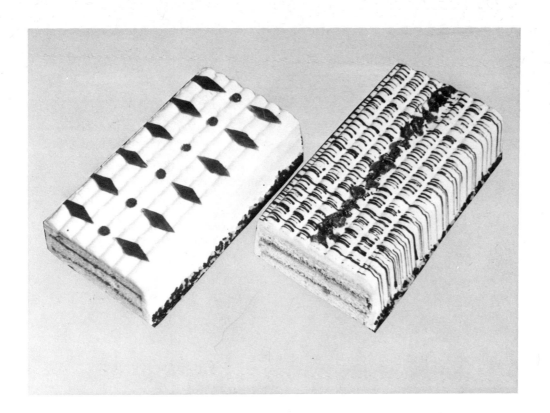

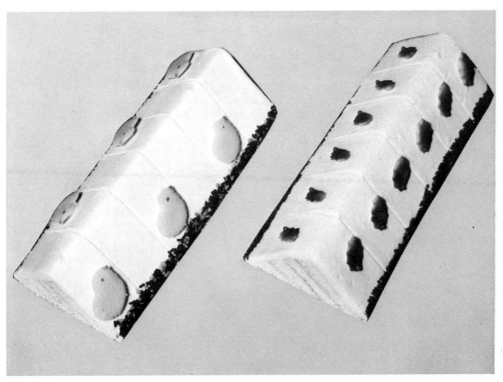

Top, Fig. 153   Bottom, Fig. 154   143

Fig. 155 – Here pattern D scraper has been used. For the interior of such shapes, a Swiss roll cut in half would be ideal: this would prevent any waste of genoese trimmings.

Both these examples are decorated with stencilled sugar motifs and sugar balls, alternately placed.

Fig. 156 – In these two examples, two differently coloured almond pastes are rolled into ropes, and arranged on top with a large one in the centre and a smaller rope each side. The whole strip is then covered with fondant. Two finishes are shown, one using crushed crystallized rose petals whilst on the other are placed small sugar balls.

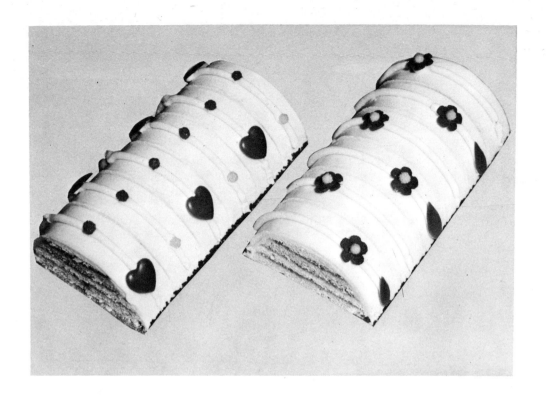

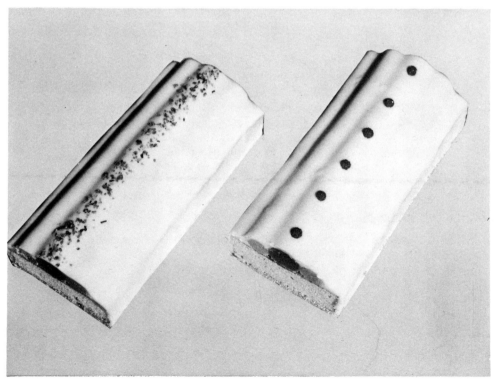

Fig. 157 – Ribbed almond paste is used to cover the slices shown. Chocolate is piped on to form the bottom edge after the divisions have been marked. In one, vermicelli-coated chocolate buttons are used for the main decoration, whilst for the other, small sugar-coated almond paste balls are used to imitate snowballs.

Fig. 158 – Both cakes are covered in plain almond paste and have a chocolate bottom edging. In each case the design is stencilled in royal icing. The example on the left first has the candle impression stencilled on and the other decoration added. Flames are piped in with yellow sugar and two prefabricated stencilled holly leaves are placed at the base of each candle. The holly spray of the second cake is touched up with green colour from a brush before the red berry is piped in with either royal icing or jelly. For a strictly commercial cake, this brushing of the leaves would not be carried out.

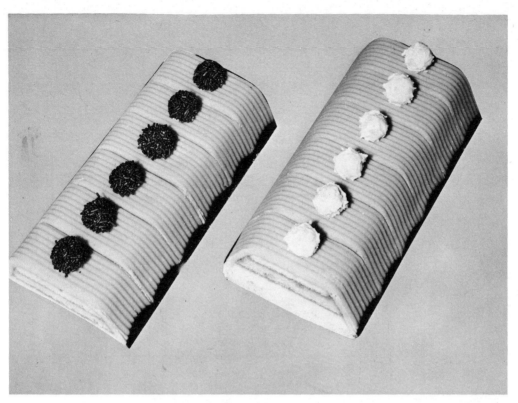

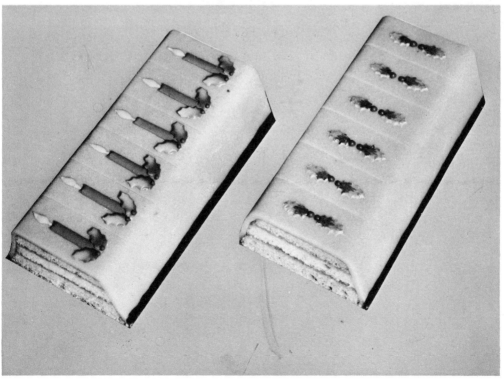

Figs. 157 and 158

Marbling

The technique known as marbling is widely used because a good decorative effect is produced with little effort. Provided the materials used are in the correct condition, anybody can use the point of a knife or a similar tool to produce an acceptable result.

However, it is the Author's experience that more time is spent on this operation than is strictly necessary. Most operators use the point of a knife at least a dozen times to effect a standard marbling pattern on to a gateau or layer cake. Such an operation can be reduced to one or two movements by the use of a special tool which consists of a number of wires evenly spaced (**Fig. 159**).

This utensil may be used to marble either the tops of gateaux (**Fig. 160**) or whole slabs of cake (**Fig. 161**). The latter is the most economical in terms of labour, since it can easily be cut into squares or rectangles and finished off into individual gateaux by masking the sides with buttercream and roasted nuts. A full-sized baking sheet of 30 in. by 18 in. would thus yield 30 gateaux 3 in. by 6 in. with a marbled top with very little effort.

Fig. 159

Fig. 160

149

Marbling with Fondant

To marble in this medium it is important for the fondant to be correctly conditioned. It must not be hot so that it sets too quickly before the marbling process, but must be sufficiently fluid so that the lines of the coloured icing merge into the rest when it is moved.

The use of the special utensil shown (**Fig. 159**) not only speeds up the whole operation, but because the wires are evenly spaced more precision is possible in the design.

The general technique is to first coat the top of the gateau or slab with conditioned fondant. Lines of a coloured fondant are then piped at intervals over the surface. The wire utensil is drawn across these lines, either in both directions or in one direction only. By drawing it at an angle to the lines more variations can be made. **Fig. 162** shows two such variations.

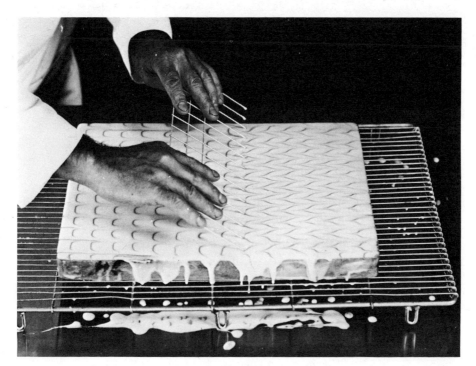

Fig. 161

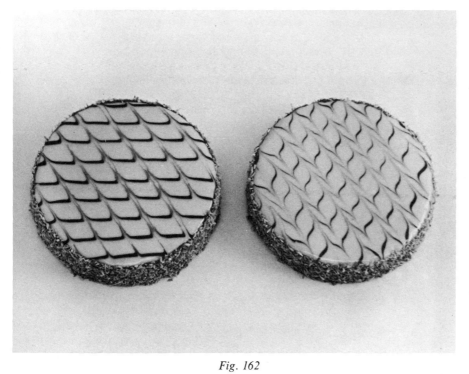

Fig. 162

151

Marbling with Filling or Buttercream

Various cream type fillings may also be used for marbling techniques.

The cream is first textured into channels into which an icing, jam, chocolate or a similar medium is piped. The wire utensil is then passed over at right angles or at an oblique angle to marble the medium into the cream.

To achieve a number of these channels a special spreader can be made consisting of a number of V or U cuts in a straight edge. The top of the slab is first coated with cream and the special spreader passed over (**Fig. 163**).

Chocolate, jam, etc. is then piped into the hollows (**Fig. 164**) and the wire passed over to give the characteristic decorative effect (**Figs. 165 and 166**).

The wire utensil is easily made by soldering several strands of wire evenly spaced on to supporting cross wires as shown.

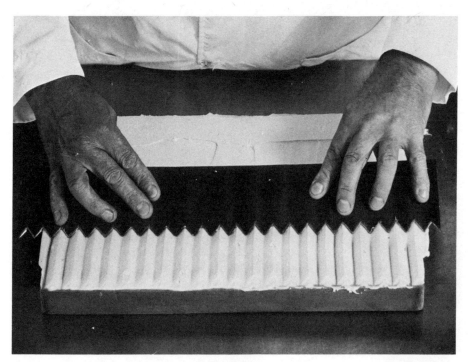

Fig. 163

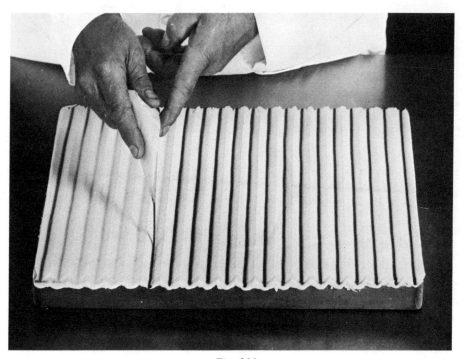

Fig. 164

153

Fig. 165

Fig. 166

154

Chapter XI

Commercial Piped Decorations

THE piped borders appear to have become rather stereotyped and one sees little imaginative use of the various tubes which are available. In this chapter are illustrated the many varied designs which can be made without any linework and with the minimum of overpiping. The use of tubes is, therefore, restricted to the following:

No. 44 or 45 – Rope
No. 32 – Point star or rope
No. 58 and 60 – Petal
No. 16 – Leaf
No. 25 – Ribbon
No. 22 – Basket
No. 2 – Writing
No. 1 – Writing
No. $\frac{1}{4}$ in. and $\frac{1}{2}$ in. – Savoy
No. 12 – Star

In addition to the above-mentioned tubes, leaf flashes are made with a bag, the end of which is cut into a 'V'-shape. The designs illustrated break themselves conveniently into four groups: shells, bulbs, rope, and scroll, and some general observations on these might be helpful.

Shells

Shells are quickly executed and make a good base for the piping on of further decoration. The piping of these should be uniform and evenly spaced around the border. With practice, it is possible to gauge whether there is sufficient space for the last four or five shells to be piped and so either to reduce or to increase the size slightly to fill up the remaining space completely.

Bulbs

In the first instance the piping of these is important to ensure that they are the same size and evenly spaced around the cake. The cake, which is to have this treatment, should be iced so that the edge is bevelled, in order for the bulbs to have a foundation on which a good round shape can be piped. After piping the bulbs, the point where the tube leaves the sugar should be flattened, either with a wet brush or a flat palette knife after the sugar has crusted.

Ropes

These can be made for tubes varying in thickness from $\frac{1}{4}$ in. to $\frac{5}{8}$ in. either with the rope or star impression or left plain. The only difficult point to watch is where the two ends meet and have to join.

Scrolls

These are probably the most popular types of border to be found in a cake. The design is quickly applied and, when well done, looks most attractive. Both the tapered and boat-shaped scrolls are illustrated. There are three movements of the tube which can bring about these shapes. The usual way is perhaps spiralling the icing but the tube can be moved from side to side and also in a jabbing motion.

Overpiping

If overpiping is required, a proper foundation should be made by firstly using a No. 4 or 3 tube over the basic pattern. The design may be subsequently built up using a No. 2 tube and then a No. 1 tube if desired.

Fig. 167 – Here the use of a petal tube is demonstrated to put a side decoration on to a shell border, the inside of which is outlined with the same tube held in a vertical position. The wavy formation of this line may be done by moving the tube from side to side or by forcing the sugar out under pressure and quickly revolving the turntable when it forms its own wavy formation.

The first way is, of course, much slower but the sugar may be kept under perfect control and it is more suited to the beginner. With practice it is possible to graduate to the quicker method afterwards.

Fig. 167

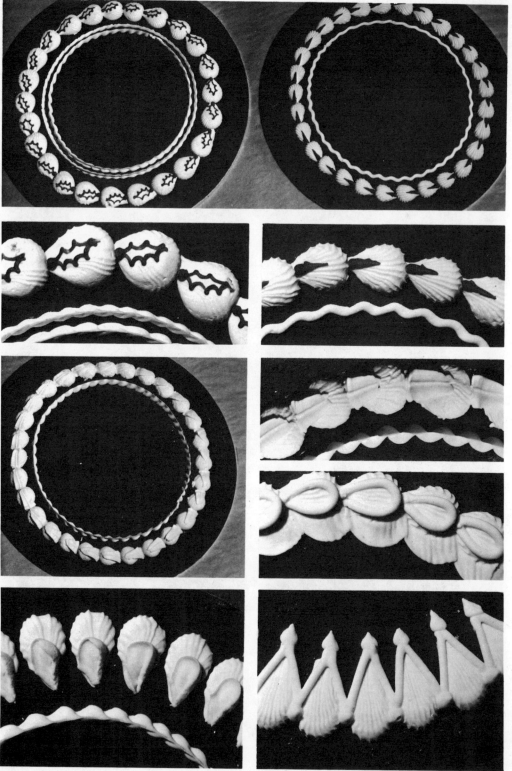

158   Fig. 168

Shell Designs

Fig. 168 – A number of different treatments to the basic shell design is shown here. The holly outline is, of course, done in green with a bright red bulb of icing piped in at its base. Flashes done with a No. 2 tube and a petal tube are seen, as well as decorative treatments.

Bulbs

Fig. 169 – Here we see a number of decorative treatments of the simple bulb design. In some of these designs – No. 44 rope, No. 22 basket, and No. 25 ribbon – tubes are used to outline the basic design, whilst a No. 2 tube is used in the overpiping.

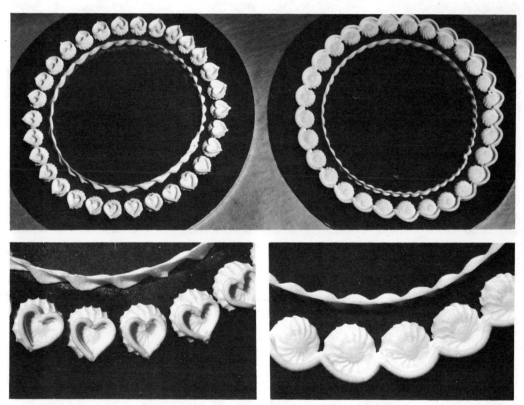

Fig. 169

160   Fig. 169

Fig. 169

Ribbon Designs
Fig. 170 – The use of the basket and ribbon tubes are depicted here as enlarged sections of borders. Additional decoration is applied with leaf, petal, star, rope, and the No. 2 tube.

Fig. 170

Scrolls

Fig. 171 – Shown in this figure are scrolls, both tapered and boat shapes, executed with a No. 44 rope tube. Overpiping is done in either a petal or a No. 2 tube.

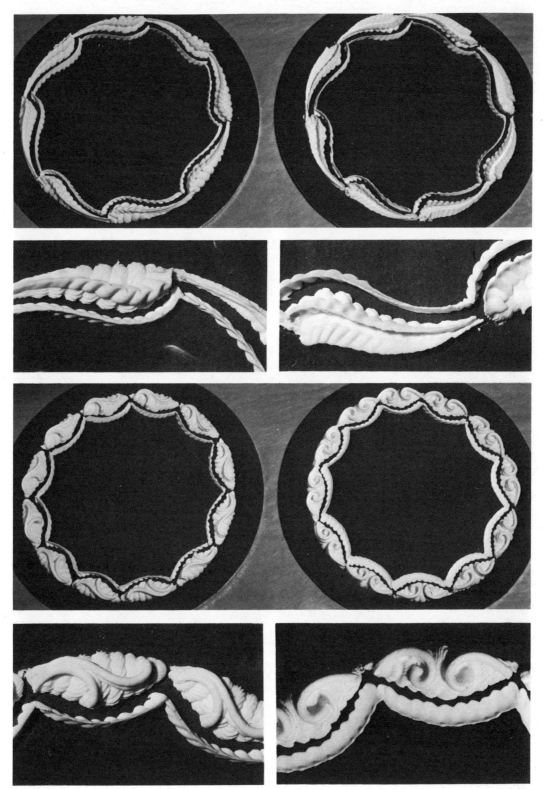

Fig. 171

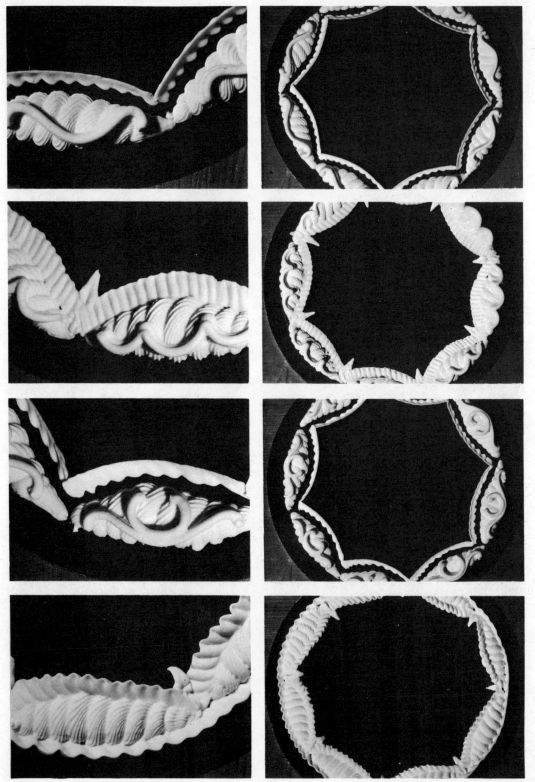

Fig. 171

Rope Borders

Fig. 172 – Borders made from a large rope or plain savoy tube are quickly applied and provide almost as much scope for decorative overpiping as the shell and bulb designs. Again the use of all the previously mentioned tubes are exploited in outlining and overpiping these rope designs.

Fig. 172

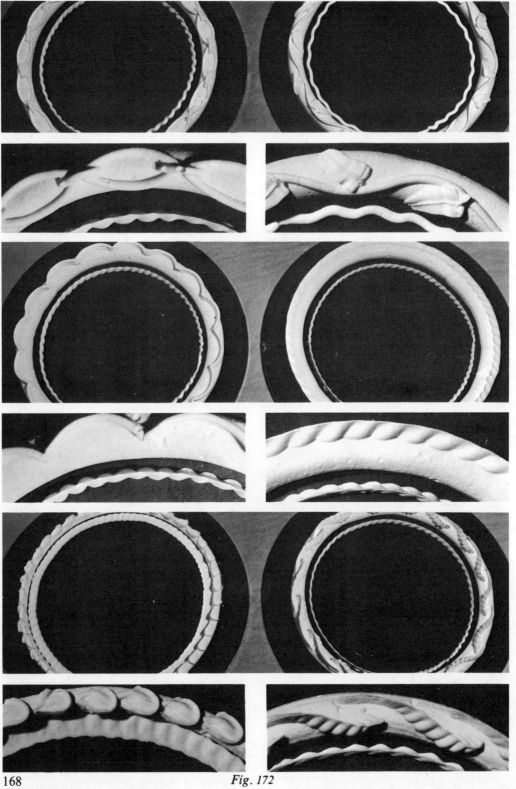

168

Fig. 172

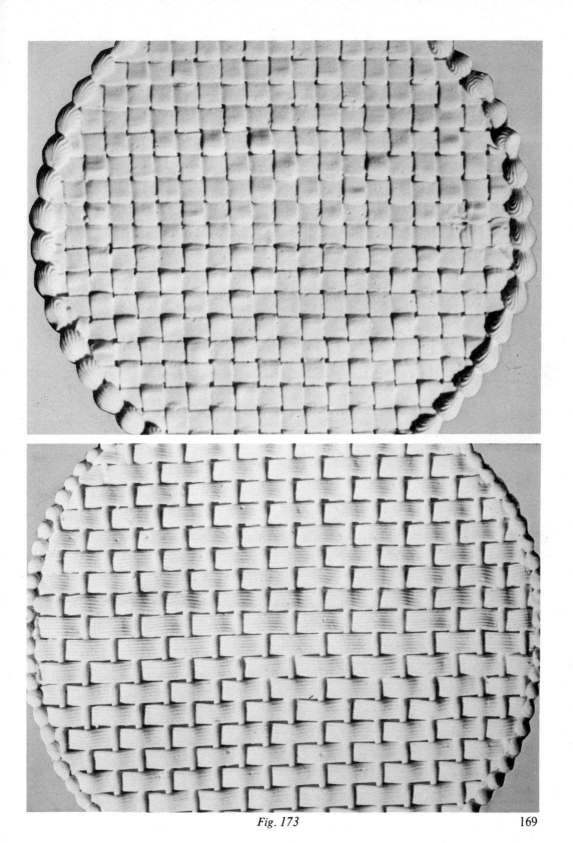

Fig. 173

169

Basket Patterns

Fig. 173 – The piping of a sugar basket pattern is illustrated. Two types are shown, one using a No. 3 writing tube in combination with a No. 22 basket tube; the other shows the use of the No. 25 ribbon tube only.

The first pattern is built up in the following stages:

1 – Short lines, the length equal to the width of the basket tube, are piped at intervals across the centre of the area.

2 – These lines are now covered with a ribbon of sugar piped from the basket tube.

3 – Lines, three times the width of the basket tube, are piped between the original ones and over the sugar ribbon.

4 – These lines are covered at each end by piping ribbons of sugar either side of the central ribbon.

5 – The basket pattern is continued from either side of this central ribbon by piping lines twice the length of the width of the basket tube from and to meet the original piped lines. These are subsequently covered with the basket tube and thus the pattern is built up.

In the second pattern, although the No. 25 ribbon tube has been used exclusively, it has been built up in the same way as previously explained.

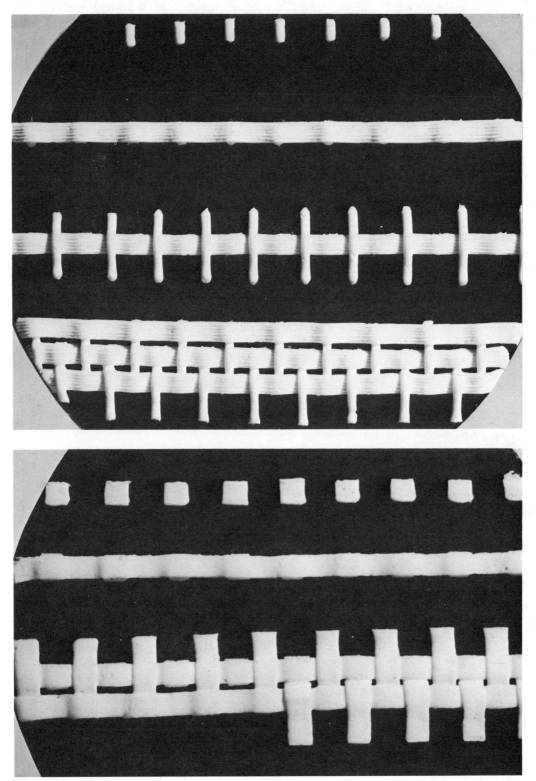

Fig. 173

Crinoline Lady

Fig. 174 – A novel form of a decorated cake is the crinoline lady and, in this figure, are shown two examples of finishes in which the leaf tube has been used. The bust is of modelled almond paste and may be purchased. This is joined with sugar to the cake base, and a wedge of textured almond paste secured to the front. The dress is now piped from one side of this to the other. For better effect two different coloured sugars are placed in the same bag.

Fig. 174   173

174   Fig. 174

Chapter XII

Decorated Run-outs

TO produce a large number of decorated cakes within a limited time some degree of prefabrication is essential. There must be adopted methods and techniques which will help the operative to decorate cakes speedily and with the minimum of manual effort. These techniques should embody jigs of all types, so that not only are the cakes executed speedily, but neatly, and with a high degree of uniformity of any particular design.

The use of run-out borders eliminates the need for the operative to spend time on piping the edge of the cake. However, the main objection in the past on the use of pre-fabricated run-outs has been the time taken to execute them. This theory is exploded by using a precision turntable and an icing aid or rest. (See fig. 1.) The method adopted by the author is used to prefabricate circular run-outs only and makes use of this equipment.

The run-out is executed on a circular, flat cake-board, which is speared on to the spike of the turntable, at its approximate centre. A disc of wax paper is then attached on to the board by means of blobs of icing at intervals around its edge. A rough measurement from the spike indicates where to start the inner and outer piped lines; this will, of course, vary with the dimensions of the cake on which the run-out is to be placed.

The Operation

Fig. 175 – Resting the tube (No. 1) and holding it rigidly against the rest and in contact with the waxed paper, commence to spin the turntable until, after one revolution, the line is joined. No attempt is made to join accurately the lines as any overlap is insignificant and commercially does not affect the final shape of the run-out. A second line is now piped, the turntable having been shifted $1\frac{1}{4}$–$1\frac{1}{2}$ in. away.

This distance could be automatically adjusted if the turntable stood between two blocks or pegs secured to a baseboard. These pegs (which could be adjustable for any size of run-out) limit the movement of the turntable, so that, when moved tight against one of the pegs, the distance of the rest from the centre of the turntable is determined accurately.

Softened sugar is then run in between the piped lines, starting inside the inner line and finishing inside the outer one. The sugar is piped out through a No. 4 tube in a spiral motion, keeping the pressure on the bag uniform and the turntable spinning at a fast speed.

Whilst all this seems tricky when thus described, it is a technique which can be acquired readily by even a beginner in cake decorating. As for the speed of execution, this needs seeing to be believed. The author was

Fig. 175

timed in one record-breaking attempt and, from picking up the piping bag to the execution of the finished run-out, took exactly 45 seconds. It is not suggested that this time is one which can be expected from an operative or that it can be done every time. A break in the line, due to an air pocket, will add at least 10

175

seconds to this figure. Also, of course, the time taken in the preparation of cementing the disc of wax paper to the cake-board and filling the bags with sugar was not taken into account. This time has been quoted, however, to show that run-outs by this method are a commercial proposition.

Kept in a warm store, hundreds of such

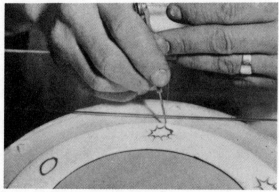

Fig. 176

run-outs could be executed at slack periods for use at times, such as Christmas. The author has kept run-outs for ten years without any deterioration.

The cake-board on which the run-out is executed, is lifted from the spike and put away in a warm place to dry. These boards can then be used to mount the cake on when the run-out is used. There is no unnecessary storage of empty boards, no drawings to worry about, and no transfer of the run-out from the drawing to the board on which it must dry. The whole operation is smooth, slick and very neat.

Decorated Run-outs

Since the execution of the run-out is so quickly done, some decoration of the run-out can be contemplated, to make it more attractive.

As the run-out sugar is applied before it has had a chance to skin, marbling at once comes to mind and most of the following examples are variations of this technique.

Firstly, to add to the attractiveness of these run-outs, the piped outline can be done in a complementary colour or chocolate. Con-

versely, if the run-out is in chocolate the outline can be done in white.

Holly Leaves

Figs. 176 and 177 – Firstly, an oval line is piped on to the run-out with green soft sugar **(Fig. 176)**. A needle is then used to pull the

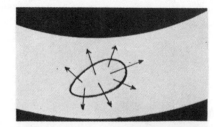

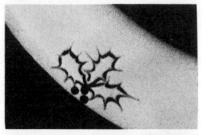

Fig. 176

green line outwards, in about seven or nine places, leaving the outline of the holly leaf in the run-out. In this figure the movement of the needle is shown. The needle should be wiped clean after each movement otherwise the green colour will contaminate the centre of the holly outline. The red berries are piped in with red softened sugar after the run-out has set and are therefore in relief.

Fig. 177 – Here are four designs in which the holly motif has been used. Notice that besides marbling various motifs, the inscription may also be piped directly on to the softened sugar of the run-out and thus is also marbled.

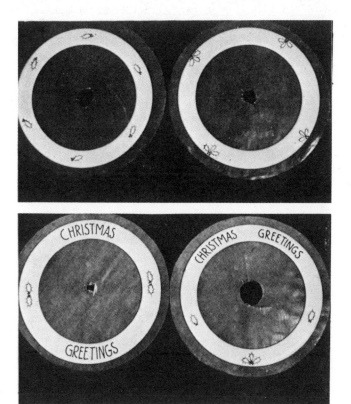

Fig. 177

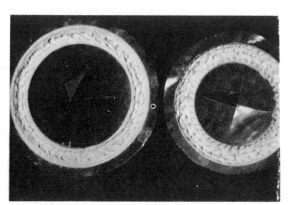

Fig. 183

Fig. 184

Other Leaves and Flowers

Fig. 178 – The flowers shown in these examples are done by first piping two circles of differently coloured sugar and pulling the lines towards the centre in eight or six places.

The leaves are similarly done by pulling the lines inwards at an angle.

An indistinct leaf impression can be created by piping a green line and pulling it in a wave motion in alternate directions. Reference to the diagrams which accompany the following figures should make these operations quite clear. Arrows show the direction of the movements of the needle in each case.

Line Marbling

Fig. 179 – Besides the attractive shape created by moving the line in a wave motion, there are many other attractive patterns which can be formed. Here we see three of them done in chocolate on a white run-out. In the first illustration only one line is marbled in a straightforward fashion. A circular movement to a thick chocolate line is shown in the second photograph, whilst the third shows the effect of the same movement on two lines.

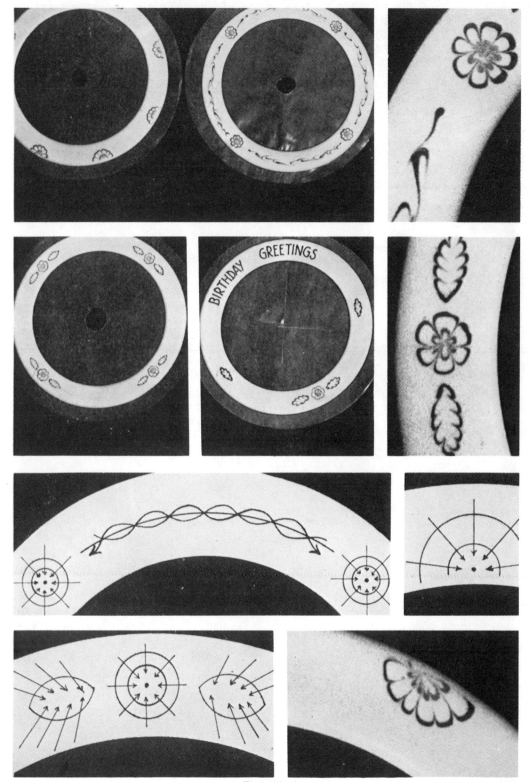

Fig. 178

179

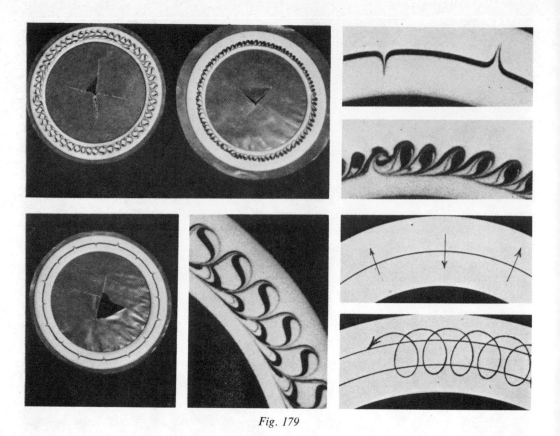

Fig. 179

Fig. 180 – Here we see the effect of using four lines of different colours for the marbling process. These are done with the spiralling movement of the needle as the diagram indicates. The last two in this figure have a circular movement imposed on the marbled design as well.

It is obvious that an endless variety of designs and colours can be thus incorporated into a run-out by this manner. There is a great danger, however, that too much movement of the coloured sugar addition to the run-out is given, which will result in complementary colours merging to give grey. Great care should be taken to see that the colours chosen blend well with each other and with the base colour of the run-out, and that movement is only given to them to a limited extent, so that the colours do not merge completely.

Two-colour Run-outs

Fig. 181 – Using two colours for the run-out also creates attractive designs, especially when the boundary between the two coloured bands of sugar are marbled into each other. In the two examples shown, one was given a scroll movement whilst in the other the needle was merely pulled inwards and outwards at the respective edges as indicated by the diagram. In each case the needle was left uncleaned between movements.

180

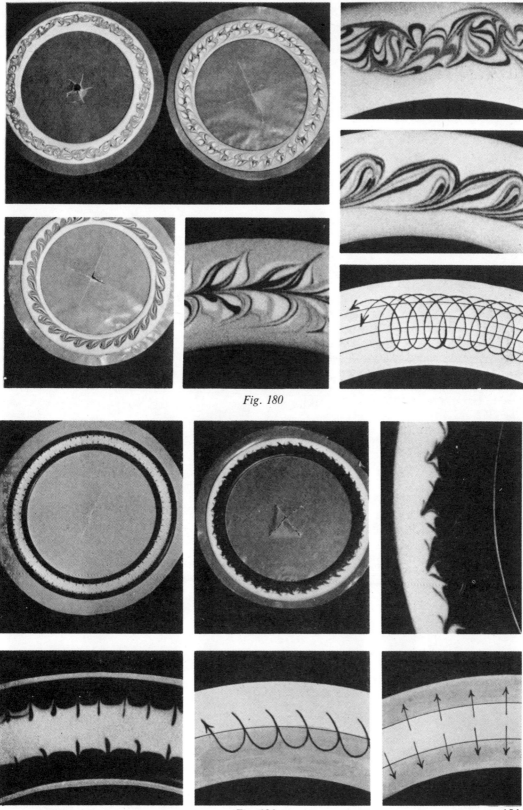

Fig. 180

Fig. 181

181

Other Motifs

Fig. 182 – Coloured sugar can also be piped on to the run-out and left without moving it with the needle, as already indicated by the inscriptions in **Fig. 177**. The Christmas tree branch and snowdrops are examples of this.

The Christmas tree and star, in the other two illustrations, are, however, marbled as the diagrams indicate.

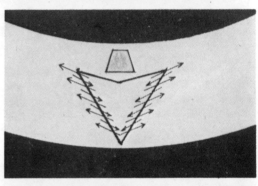

Fig. 182

182

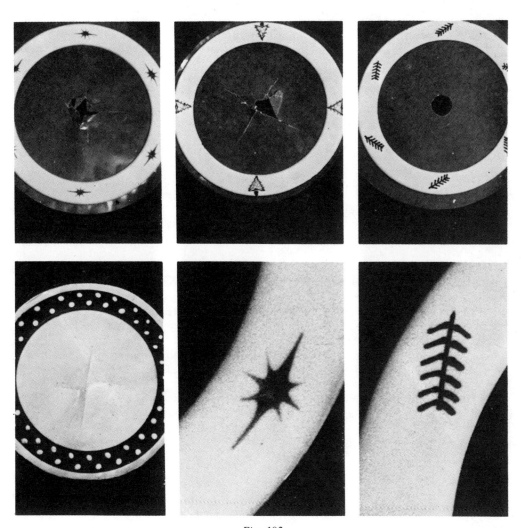

Fig. 182

183

Rough-cast Run-outs

Fig. 183 – Rough-cast sugar rims for cakes can also be made on the precision turntable. The sugar is, of course, stiff and piped out with a large tube. A palette knife is used to pull up the sugar. Two examples of this treatment are shown, one having a bold rim whilst the other is flatter and dusted with castor sugar to give it a sparkle. (See page 177.)

Decorated Plaques

Fig. 184 – Besides run-outs to encompass the rims of cakes, circular plaques can be just as easily and speedily executed. Again, no drawing is required, the size being governed by the distance that the tube is placed from the turntable spike. The plaque can be decorated and can carry either stock inscriptions or can be made for special occasions. A range of plaques suitable for Christmas and birthday cakes are shown here. (See page 177.)

Sugar Dusted and Stippled Run-outs

Fig. 185 – Using the techniques already described we may also make run-outs even without the two stiff retaining lines. The sugar for this needs to be slightly stiffer than that required for marbled run-outs. In fact, it needs to be only just soft enough for the marks of the tube or bag to disappear when the board on which the run-out is placed is gently tapped.

Naturally, the edge of such a run-out is rougher than one which is piped, but if the run-out is dredged, or stippled with royal icing afterwards, it is indistinguishable from its piped counterpart. The production of these run-outs is illustrated in the figure. Notice how prominent the lines of softened sugar are, after piping in. These disappear after the board is gently tapped and the castor sugar may be dredged on. Omitting the piping of the two retaining lines speeds up the production of these to the point that they can be accomplished in under 30 seconds. Careful scrutiny of these two photographs will prove how perfect such run-outs can be.

With the stippled variety, softened royal icing is applied to the dry undusted run-out, with a clean sponge. The sugar can either be of the same colour, or of another blending colour, to form a very attractive finish. Since the stippling effect is irregular, any irregularities in the edge of such a run-out are completely masked.

The use of all these run-outs is explained and illustrated in Chapters XIII and XIV.

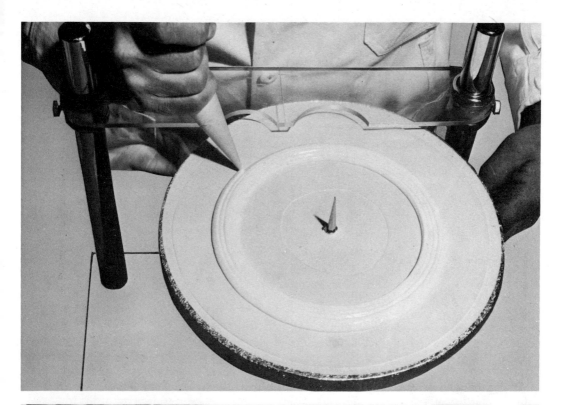

Fig. 185   185

Chapter XIII

Christmas Cakes – Use of Run-outs

A QUICK, commercial method of decorating cakes using the circular run-out described previously is now given.

The cake is first almond-pasted, fixed to the cake-board and given a fairly thick coating of sugar on the sides only. For a quick commercial job, one application will suffice if the sugar has been properly made and conditioned: two coats would be necessary for an absolutely smooth side. When the side coat is dry, the circular run-out is placed and cemented to the top with sugar. Soft sugar is now run in, using the turntable and rest and spinning the cake at a fast speed. The sugar is run in, starting at the centre and gradually spiralling to the inside circle of the run-out, using the same coloured icing as the side coat. This action is illustrated in **Fig. 186**.

Executed in this way, the top coating is applied in a fraction of the time normally taken to coat a cake and the results bear comparison with a conventionally iced cake. Moreover, the soft top enables the flowers, leaves or any other design to be executed in marbling as well as the inscription.

The bottom splay may be run-out and marbled also to match the top. Sugar is run on to the cake-board in the same manner as with the top. No outside retaining line is necessary for, if the sugar is of the correct consistency, it can be spread against the edge of the board and will not flow over.

The second illustration in this figure shows the white dots being applied to the bottom splay so that it matches the top run-out, thus completing a very effective design.

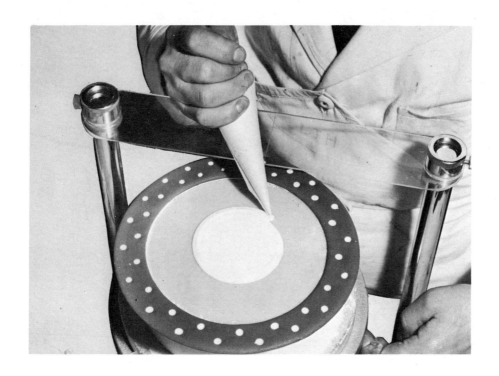

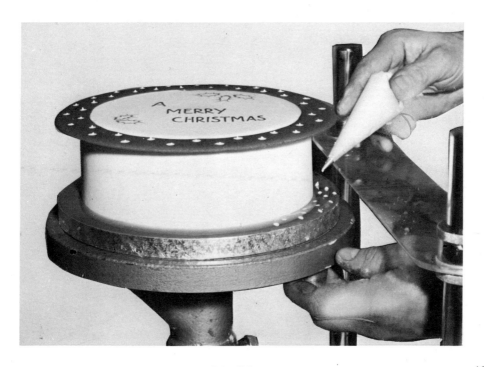

Fig. 186   187

Fig. 187 – Two examples on the use of run-outs by these methods for Christmas cakes. Additional decorations are an almond-paste log, plastic robin, and holly sprays.

The great advantage with this method is the speed with which cakes can be decorated. Adopting these methods any worker, after a little practice, could decorate a cake speedily and neatly. No precision linework is required and very little skill is required to coat the sides of the cake. The main skill lies in the almond-pasting of the cake to ensure a level top.

Fig. 187

189

Sugar and Almond Paste Tops

Fig. 188 – Another method for a quick finish to a cake is the use of prefabricated, sugar, or almond, paste discs. The paste is first pinned out to approximately $\frac{1}{8}$ in., then cut into circles of such diameter that each overlaps the cake by approximately $\frac{3}{8}$ in. This cutting out is best done by a specially made cutter, since the circles require to be perfect in shape. The paste should be pinned out on to a board and, after cutting, only the scraps are removed, discs being left on the board until set. If a cutter is not available, a disc of metal, or a cake-board, can be used and the discs cut out with a sharp-pointed knife. When set, these discs are cemented to the top of the almond-pasted cakes which have had their sides only coated. The Christmas cake illustrated in the figure is finished off in the following manner:

A line of stiff sugar is piped about 1 in. from the edge and made rough with the palette knife. The remaining edge is now coated with soft sugar, running in from the rough cast ridge to and over the edge, so that it begins to drip down to form imitation icicles. Holly leaves are marbled into this rim, as illustrated in the previous chapter **(Fig. 176)**. By the time this has been done, the soft sugar has begun to set and no more will drop on to the bottom splay, which can now be coated with stiff sugar and rough casted, or soft sugar marbled to match the top. A decorated plaque in the centre of the top and the artificial holly leaf on the bottom splay complete the design.

Fig. 188

191

Colour-sprayed Cakes

In the next illustrations the air-brush has been used to colour the top and lower borders of the cake and a simple bottom border design is applied. **Fig. 189** shows how these are done.

Colour is applied to the top edge of the cake by holding the air-brush in such a way that the spray is directed away from the cake centre in the manner shown. The cake is revolved a few times until the colour is applied in sufficient strength. Now the bottom is sprayed, directing the air-brush to give a band of colour between the splay and the cake. White sugar is piped on to this band of colour and is pulled out, using the points of a pair of scissors or a similar tool, as illustrated in the figure. In the following illustrations the colours used are either chocolate or blue:

Fig. 190 – This shows the use of a white, stippled run-out with a chocolate-colour-sprayed cake. The inscription is executed in a No. 3 flattened tube applied directly to the cake with an embellishment carried out in a No. 0 tube. On an area of roughened icing is placed an almond-paste log and a plastic robin, whilst the side is decorated with a green holly outline, piped with a No. 0 tube, and red berries. The design is simple, quickly executed and effective.

Fig. 191 – Use of prefabricated, sugar, stencilled run-out holly leaves, and almond-paste inscription, has been made in this cake. Two circles of orange colour are sprayed on to the top and two candles are piped in red sugar to meet them. A bright yellow flame is now piped over the orange circle with a No. 1 tube. The holly leaves of different sizes are arranged as a spray at the foot of the candles and red berries piped in. A smaller spray of holly is put on the side. The border run-out used in this, and the other examples, is of the sugar dredged variety.

Fig. 189

Fig. 192 – Here stencils are used to design the Christmas tree. A stencil is first used to colour seven circles of orange on to the top surface. Superimposed is another stencil of a tree shape with a tub. This is stencilled with chocolate-coloured-icing whilst the green icing is used for the tree shape.

White lines of royal icing from a No. 2 tube are piped from the green tree shape to meet the orange circles and, from here, yellow flames are piped over the orange colour. The tree provides a useful motif for carrying the inscription which is piped on with a No. 1 tube. The side is decorated by first spraying circles of orange and setting up a red sugar-piped, or stencilled, run-out candle with a yellow flame piped over the orange circle. Two run-out holly leaves with berries complete this side design.

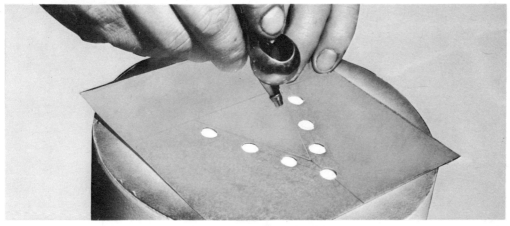

Fig. 192   195

Fig. 193 – Another arrangement of the pre-fabricated decorations is shown in this figure. Here we see the use of the stencilled run-out robin. This is set into position with a bulb of stiff icing: then the legs are piped in with a light, chocolate-brown-coloured icing.

Fig. 194 – A lantern is stencilled in this example. The stencilling is done in chocolate over an area which has been previously sprayed in orange colour. Again use has been made of prefabricated holly leaves and inscription. The border in this case is piped directly on to the edge of the cake, the inside pulled out with the scissors to match the bottom border and then roughened up with a knife.

Fig. 195 – Shells piped in white and overpiped in a green holly leaf outline form the border decoration of this cake. Bright red sugar bulbs are piped in for the berries. The combination of the white, green, and red against the sprayed background is very effective. Sugar-edged stencilled plaques and an artificial holly spray complete the design of this cake.

Fig. 196 – Here the top border consists of a white almond-paste ring, which is stippled over with white royal icing after assembly. The paste rings may be made by rolling out a rope about $\frac{1}{2}$ in. in diameter and either placing it directly on to the cake edge, or around a template to set. Whether pre-fabricated or directly applied it should first set before it is subjected to the icing rough-casting. These borders are easily and quickly prefabricated and may be stored and used as a border decoration by unskilled labour, as and when occasion demands.

Before applying this almond-paste rope, the top centre of this cake is sprayed blue and over this is spread a thin layer of neutral piping jelly. This produces the effect of a pool which has iced over. The illusion is emphasized now by running in softened white icing from the edge of the cake to terminate in an oval surrounding the pool. The inscription is written first in white, then overpiped orange over this pool. Stencilled run-out holly leaves are used to add interest and colour and the bottom splay is roughened to match the top.

Fig. 197 – The border here has been first piped then covered with softened sugar. The piping is done with a No. 3 or 4 tube, using the precision turntable and icing aid. Rings of sugar are piped on top of each other to form a wall which is made to turn over at the top and project over the cake. When this has been covered with soft sugar, icicles are piped down, using a No. 2 tube. Whilst the piping of this border can be made commercial in the hands of a skilled decorator, the piping of the icicles cannot and it would be better from this point of view to allow the soft sugar to drip off and form their own icicles as described in **Fig. 188**.

A sugar band is piped inside this border and roughened to match the bottom decoration. Holly leaves, made from stencilled sugar, adorn the two borders. A cherry snowball, with robin and the inscription, completes the design.

Fig. 198 – The border in this illustration is done with a $\frac{5}{8}$ in. savoy tube piped directly to the edge of the cake. Stencilled holly leaves are used to cover up the join and also decorate both borders. The scene is executed firstly by piping the fence posts in chocolate colour and, when set, covering the top with white softened sugar, to imitate snow. Into this is placed a stencilled robin.

The inscription is written on in the base colour and overpiped in chocolate colour.

Colour Schemes

For the majority of cakes illustrated in this and the next chapter, the bases are coloured in a cream or stone colour (egg yellow plus a very small spot of french pink) since, being neutral, this blends well with any other colour which may be applied. Chocolate, red, green, orange, and blue, may all be used with good effect provided the quantities and strengths are correct. On this last-mentioned point some comment might be helpful. It is often necessary to use icing which is very strongly coloured, especially for the red holly berries, inscriptions, etc. If this is piped directly on to the base, there is a danger of the colour seeping from the decoration to the base, which will act like blotting-paper and produce an unsightly patch in the vicinity of the decoration. Whenever possible, and particularly in the case of inscriptions, the outline should first be done in the base colour then overpiped in the vivid colour. With holly berries, these can be piped on to the holly leaf or outline.

Being very small, it should form a perfect sphere, if the consistency is right, and this, too, will prevent the blotting-paper action of the base coating.

The cakes illustrated and described in this chapter are all quickly and easily decorated without any intricate linework. They rely on the use of colour and perhaps unusual treatments for their effect. The use of prefabricated décor helps to make these cakes more of a commercial proposition and is especially recommended for occasions such as Christmas.

An endeavour has been made in this chapter to depart from the use of conventional forms of piped designs, but at the same time to show a very comprehensive range not only of borders but inscriptions and treatments, some perhaps unusual.

By permutating all the designs shown, a very wide range is possible but this must be the prerogative of the confectioner.

202

Chapter XIV

Birthday Cakes

MUCH of what is written and illustrated in the previous chapter can equally apply to birthday cakes. Obviously, by using the piped border designs of Chapter XI, a very wide range of decoration may be adapted. This, however, the reader can achieve for himself and so the author aims here to give more unusual treatments for the decoration of this traditional cake.

With the exception of one, all the cakes are strictly commercial, decorated with the minimum of linework and employing pre-fabricated decorations.

Fig. 199 – This cake is piped in the conventional manner, using only the shell and leaf tube for the border decoration and a No. 0 tube for the inscription. For effect, the leaf bag has two different coloured icings placed in prior to piping out the design. Prefabricated sugar stencilled flowers and leaves are used for the top and sides.

Fig. 200 – In this example the top edge and lower splay is sprayed pink before the piped decoration is added. This is embellished with prefabricated sugar flower and leaf forms and also sugar balls. The inscription is carried on a pink, almond-paste disc. As this illustration is in black and white, the sprayed colour does not show up very well. This colour should not be very strong and thus be carefully applied. In the examples shown, in which colour is applied, french pink is used and sprayed on to a pale cream base colour.

Top, Fig. 199   Bottom, Fig. 200   205

Fig. 201 – A stippled run-out is used for the border in this cake. The run-out is prepared in pink sugar as described in Chapter XII, and, when set, is stippled in white icing. The splay is similarly executed. Stencilled flower petals are assembled together with green sugar stems and leaves to form the floral spray on the top and this encompasses the almond-paste plaque which carries the inscription.

The use of pink satin ribbon eliminates the need for side decoration and its attractive form is ideally suited for use in birthday cakes.

Fig. 202 – The sugar-dredged type of run-out is used in this example. This is executed in yellow and the bottom splay is similarly done to match. On this splay stencilled white flowers and green leaves are arranged in six places. The top is simply decorated with the inscription and three candles. Again satin ribbon is used, this time green, to finish off this quickly executed and strictly commercial cake. This cake, too, is tinted with the colour spray around the top and bottom edges.

Fig. 203 – The use of a scraper to impart a rib or similar decoration to the side of a cake is not new and scrapers can be purchased with different patterns cut into them. What is not always appreciated is that not only may a side decoration be applied in this way but the bottom decoration as well. Also, that such scrapers may be easily cut by the operative from any plastic scraper, using a craft tool. For rigidity, however, the author prefers the use of a plastic set-square which is just as easy to cut.

To smooth the cut edges afterwards a piece of fine emery paper may be used.

The scraper is used in the following way. Royal icing is applied liberally to the sides of a cake to which one coating of sugar has already been given and allowed to harden. A rope of sugar is piped around the lower edge between the cake and the board. The scraper is held in position and pressed against the

Fig. 203

cake as it is made to revolve on the turntable. This action is illustrated in **Fig. 203**. The space between the outside edge of this rope and the edge of the board may be covered with a coating of sugar when the rope has set or cleaned up to show the silver board.

Since a rope has been used for the bottom border, a similar decoration should be used for the top. This may be piped directly on to the top edge, the use of an almond or sugar-paste ring, or the rope ring may be piped out

on to wax paper and left to harden before transferring it to the cake. This last method has certain advantages in that it may be piped with precision, using the icing aid and turntable, and it can be made to project beyond the cake edge in the same way as a run-out and so give the cake a better proportion.

As with all prefabricated decorations, its use enables cakes to be speedily decorated even by the novice decorator. In the following two figures these sugar rings are used:

Fig. 204 – Before applying the rope border decoration, this cake is colour-sprayed around the top and lower rope border. A small leaf tube is used to outline the borders and this is the only piped work actually done on the cake. Prefabricated stencilled flowers, leaves, and inscriptions, complete this strictly commercial design.

Fig. 205 – This same rope design can be used to produce a more elaborately decorated cake. Even so, the linework is reduced to the minimum. The flowers decorating the borders are piped and are used in conjunction with stencilled sugar leaves. The main flower is made from six stencilled petal shapes and assembled with a yellow flower centre.

From this example it must be obvious that a very wide variation of this basic rope design is possible and would look equally as effective.

A birthday cake is a personal item of confectionery and, therefore, the wishes of the customer with regard to the inscription or occasion should be treated with respect. For example, if the cake is to be given to a child, make sure that the child can read for himself the inscription. Make sure it is bold, large and in a form that is easily understood – not script or Olde English. Ensure the inscription contains the name of the child and that the decoration includes room for candles.

If the cake is for a Twenty-first Birthday, it is desirable to have a key motif and the number 21 in the inscription. In short, make the cake fit the occasion and you will always have a satisfied customer.

Chapter XV

Wedding Cakes

BEFORE illustrating types of design which are applicable to wedding cakes, some explanation on the preparation of the base might be helpful.

Firstly, the cake itself should be mature, that is, it has been made and stored for a few weeks to mellow. During this period rum or brandy could have been rubbed in from time to time.

The cake base is coated with boiled apricot puree and covered with a good quality almond-paste. This paste should be specially prepared and of the right consistency. For tiered cakes it is important that the paste be stiff enough to withstand the pressures put upon it by the other tiers. A multi-tiered cake might require a boiled almond-paste or even supporting pillars fixed inside the cake bases of the lower tiers.

A thin coating of boiled fondant is now applied to the top and sides of the base. This is to seal the almond-paste and to prevent it from discolouring the icing by its natural oil.

The cake is mounted on to an appropriate size cake-board and iced with royal icing. Two coats should give a good commercial finish but, if a perfect surface is required, three coats must be given. If part of the edge is to be left showing in the design, the coating should be done in two stages. The top should be first coated and the edge trimmed by running a knife against the hardened side coat. When this has set and is hard, the side coat is applied and trimmed off against the hard top surface. In this way a perfect right-angled edge can be built up.

Pillars

These can be purchased made in gum paste, plaster, plastic, paper, card, or metal. The choice of material is relatively unimportant, except from the point of view of appearance. With the possible exception of the card pillars they are all usually capable of standing up to the weight required. However, it should be realized that the weight concentrated on the small area of cake which is in contact with the pillar may be considerable. For example, if we assume a weight of 10 lb. for the middle and top tiers (which is not excessive) distributed on three pillars the dimensions of which are approximately $1\frac{1}{2}$ in. at the base, the pressure exerted on the cake in these three areas is $\frac{10}{3} = 3\frac{1}{3}$ lb., or nearly 2 lb. per square inch! Think how this pressure increases by extending the number of tiers to 4, 5, or even 6! Under such conditions, supports must either be built into the cake or a plate used to spread the weight evenly over the top surface. In this connection a simple, double-thick strawboard could be used. Some pillars are already made attached to such a plate.

Proportions

Although the size and weights of cakes vary, the proportions of the cake and its board should remain the same. The tiered cake should represent a pyramid of pleasing proportions. For a two-tier cake there is little or no problem, for providing the top tier is about a half or two-thirds smaller than the bottom tier, the proportions would look all right. The multiple-tiered cake, however, presents us with a much greater problem, for not only has each cake to look in proportion in itself but also with the other cakes in the tier. In **Fig. 206** a three-tier cake is shown drawn to scale from the following measurements:

	Cake Diameter	Cake Height	Board Diameter
Bottom tier	$10\frac{1}{2}$ in.	$3\frac{1}{2}$ in.	15 in.
Middle tier	8 in.	3 in.	11 in.
Top tier..	$5\frac{1}{2}$ in.	$2\frac{1}{2}$ in.	8 in.
Thickness of board ..	$= \frac{3}{8}$ in.		
Height of pillar ..	$= 2\frac{1}{4}$ in.		
Height of vase of flowers ..	$= 8$ in.		

It will be noticed that the cake height reduces as well as the diameter, so as to keep

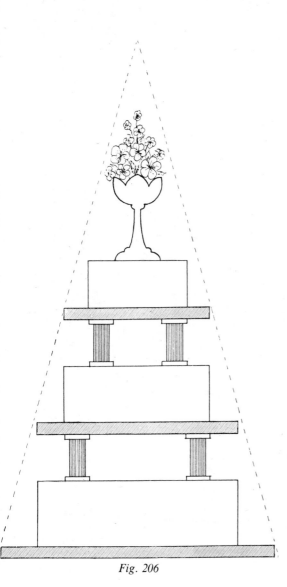

Fig. 206

each individual cake well proportioned. Also the splay on the boards are reduced, the bottom tier having 2 in., middle $1\frac{1}{2}$ in. whilst the top is $1\frac{1}{4}$ in. Of course, the size of the boards are to some extent determined by what is available and since sizes increase by 1 in. only, we have to use boards to the nearest inch. The middle board in the figure, for example, would be best $11\frac{1}{2}$ in., but this is unobtainable.

Naturally, the sizes and weights of each tier and of the cake as a whole is a matter for the confectioner, and may vary greatly from those given, but, provided the same kind of proportions prevail, the cake can be flatter, taller, or of different sizes and still appear to be balanced.

Decorations

Traditionally the use of artificial decoration is expected on wedding cakes and, if used with thought and discretion, they can add greatly to the overall design. The indiscriminate use of such decoration, however, can not only spoil an otherwise good design but be wasteful as well. When designing a cake, thought ought to be given to the type of decoration required and this made to blend in with the design. For example, if filled shoes are to adorn the edge of the cake, then space ought to be left in the border for it.

However attractive artificial decorations can be there are a few decorations which may be made out of sugar and two of these are here shown.

Bells

Fig. 207 – Sugar bells may be made as follows: Using a No. 3 or 4 tube, pipe out on greaseproof paper the bell shape, first by making a bulb, lifting the tube slightly and then forcing out the sugar and at the same time gently lifting the tube until the bell shape is made. Then the pressure is released and the tube lifted off. These are piped out evenly in rows on to greaseproof paper and, when the tray is full, it is placed under top heat for the top of the bell shape to harden. After about half an hour the bells are removed from the heat, carefully picked up and the

211

Top, Fig. 207   *Bottom, Fig. 208*

soft sugar inside scooped out with a pointed knife or a scissor blade. The hollow bells are left to dry out completely.

When used, the clanger is piped in with a No. 1 or 2 tube.

Rings

Fig. 208 – The piping and running out of rings to surmount the edge of cakes used to be a lengthy and tedious job. The use of a stencil eliminates this and in the following we explain how these rings can be made accurately and speedily by this method.

Firstly, a stencil has to be cut with a number of impressions about $\frac{1}{4}$ in. wide and the length three-quarters of the circumference of the circle desired. In any event a smooth wood roller of this diameter must also be made to act as a template, on which the rings can dry. The stencil is laid on to a strip of wax paper and white softened royal icing spread through. This strip is now wrapped around the appropriate roller, the ends secured with a spot of sugar and transferred to the drying cupboard, where it is left until set.

When dry, the wax paper is carefully cut and the whole strip slipped off the roller. By curling the paper away from the inside, the rings are released and can now be applied to the cake.

In the limited number of illustrations which follow the use of decorations is explained.

Run-out Borders

Fig. 209 – This three-tier cake utilizes the rings stencilled previously described. The run-out border is cut away at four places to accommodate the rings. Simple piped flower forms adorn the sides whilst artificial flowers and fern are used on the borders of each tier.

Fig. 209

213

Fig. 210

Fig. 210 – This two-tier cake is designed on similar lines but is decorated in the exhibition style. The run-outs, which are used for both top and bottom borders, are applied in four pieces, using an artificial flower and two silver leaves to cover over the join. Artificial flowers are also used for the spray on the sides. The monogram is a run-out sugar piece, painted silver.

Traditional

Fig. 211 – This three-tier cake is traditionally piped in a scroll design and, in addition to the liberal use of artificial decorations, the sugar bells are also used. Notice here how the design has been broken, to allow the slipper to stand on the textured edge of the cake.

214

Fig. 211

Fig. 212 – Square wedding cakes are becoming increasingly popular and here we see one which is typical. The monogram in this example is piped on, but this could perhaps be more easily stencilled. This is an example of a commercial cake decorated with the minimum of linework. Notice the attractive use of a silver vase and fresh flowers instead of the more usual artificial ornament.

Fig. 212

215

Fig. 213 – Heart shapes are also popular, and in this example we see a three-tier cake in this shape. Instead of outlining the bulb border in graduated piping, a basket tube is first used and this overpiped with a leaf-cut paper bag. The flashes on top of the overpiped bulbs are also done with a paper bag cut into a 'Vee'. A spray of white sugar-piped roses are set against a background of silver leaves at three places around the sides and sugar bells are again shown. The middle-tier contains the monogram in a stencilled form.

216   *Fig. 213*

Fig. 214 – A wedding cake is perhaps the most personal item of confectionery which can be made and since, for most people, it is something that does not recur its thought is treasured throughout their married life. Many couples, therefore, ask for certain personal motifs and themes to be incorporated into their cakes. In this last figure, many motifs were incorporated at the wishes of the prospective married couple. The bride was especially interested in music and the theatre and so the masks of comedy and tragedy were made in run-out sugar and placed on one side of each tier. A few bars of music with sugar bells placed over formed the main decoration on the sides and effectively provided the music theme of the bride.

The groom was a graduate at Trinity College, Cambridge, and this was depicted with the use of a shield of Trinity College executed in sugar and a sugar model of Trinity fountain, which was used instead of the traditional ornament to surmount the top. This sugar model was executed in gum-paste and sugar run-outs and was mainly the work of one of the students of the Cambridge school to whom it was given as an exercise in modelling. Lastly, the bride and groom's initials were linked by the monogram, done in run-out sugar. The rest of the decoration was simple with overpiped bulbs and linework reduced to the minimum.

Cutting a Wedge

Sometimes a wedge is required to be cut. Usually this must be considered with the design which should be spaced accordingly. The wedge is cut out after the cake is almond-pasted and given its final coating of sugar. Greaseproof paper is cut and inserted to line the interior cut surface of the cake, whilst a similar piece lines the corresponding surface of the wedge. Silk satin ribbon is now wrapped around the wedge and a bow made. The wedge is replaced and the final decoration applied. For a better finish the cake may be given another thin final coating of sugar on top after the wedge has been replaced. This will cover the cut and yet allow the wedge to be removed easily when required.

The colour of the traditional wedding cake is white but, in order to achieve the illusion of a pure white, it is necessary to add a little blue colour. This pure white should, however, be relieved with the intelligent use of decorations, particularly those such as pink or purple heather and flowers with yellow centres, and, of course, silver or gold. There is a trend today to break with the traditional and occasionally there is a demand for a coloured cake. Usually the colour is required to match the bridesmaids' dresses. Such cakes must be executed in the palest of tints and the use of strong colours avoided.

217

Fig. 214

Stencilled Borders

The use of waxed parchment stencils has one serious drawback which mitigates against their use to produce large run-outs of sufficient thickness to provide the strength necessary for commercial production.

However, we can overcome this deficiency by cutting stencils in cardboard of the required thickness and making them waterproof afterwards by giving them three coatings of French or button polish, allowing each coat to dry before applying the next. The polish should be applied carefully with a small paint brush so that it soaks into the cut impressions as well as the surface. Such stencils are comparatively easy to cut. All that is necessary is for the design to be drawn and then cut out with a sharp craft blade. These stencils have good resistance to moisture and can be left to soak in water to dissolve off excess sugar for cleaning after use.

Stiff royal icing is used. Run-out sugar would flow too much, causing the pieces to be misshapen.

Circular Bands

The cake shown in **Fig. 215** was decorated using eight sugar bands for the border and a lily centrepiece of six petals. The bands were stippled to give it a roughcast effect and the base executed likewise. The stippling is a useful way to camouflage the slightly rough edge caused when the stencil is lifted away from the sugar impressions. This operation is shown in **Fig. 216** and consists of dabbing a sponge, which has been previously placed into royal icing, on to the dry surface of the stencilled shape.

The detailed method of producing these bands is as follows.

Cut impressions into the strip of cardboard. The size of each will depend upon the size of the cake to be decorated. The number will depend upon the chosen design and their spacing. In **Fig. 215** the size chosen allows for silver leaves to be used between each decoration. The shape required depends on the shape of the cake, i.e. square or circular, and the size of the template around which the stencilled pieces are placed to set.

Fig. 215

Fig. 216

220

As a guide the diagram in **Fig. 217** may help.

A = Width   B = Circumference   C = Radius of cake.

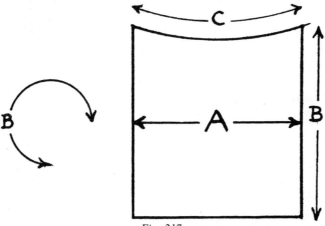

Fig. 217

Obviously for square cakes C would be straight. After the stencil has been cut and made waterproof, the rod or tube over which the stencilled pieces are to be left to set must be prepared. For the bands illustrated a tube of $\frac{3}{4}$ in. was used. This must be covered with a sheet of waxed tissue, the ends of which should be secured with a line piped in royal icing. The waxed tissue wrapping must be loose and able to slide easily off the rod or tube.

Next, strips of waxed tissue of an appropriate size ($1\frac{1}{4}$ times the circumference of the tube) are laid into the corner of a baking tray and the bottom edge secured to the tray with a line piped in royal icing. This is so that the waxed tissue is not disturbed when the stencil is removed.

The stencil is then laid on to the waxed paper and stiff royal icing spread through with a palette knife (**Fig. 218**). A long straight edge is then used to spread an even coating of icing and to remove the excess (**Fig. 219**).

The use of a tray with upturned edges enables the stencil to remain fixed in one position during the stencilling operation. This is essential since both hands are required for the final spreading of the sugar (**Fig. 219**).

The stencil mat is now removed from the same edge used for securing the waxed tissue to the tray with icing (**Fig. 220**). In this way a fairly clean edge is made on release and the impressions undisturbed as the mat is lifted away. Now the row of stencilled bands are wrapped round the rod or tube and again secured to the wax tissue covering with royal icing (**Fig. 221**). This is best done in the following stages:

1 – Pipe a line of icing on to the prepared tube and to this secure one edge of the waxed tissue containing the stencilled pieces.

2 – Turn the tube to wrap the stencilled pieces around it.

3 – Secure the other edge of the waxed paper containing the stencilled pieces with royal icing.

4 – Support the tubes from each end. For this purpose slabs of polystyrene into which Vees are cut, can be used.

Fig. 222 shows these stencilled bands placed on to the tube to set, together with the stencil used.

Figs. 223 and 224 show another set of eight plain bands on which a line has been piped at each end prior to their being placed on the tube to set. To remove these bands first slide the whole wrap of wax tissue off the tube (**Fig. 225**). Using a sharp knife cut through the sleeve of wax tissue and carefully remove the individual bands (**Fig. 226**).

To fix these decorative pieces, it only requires a line of piping on each side before placing them into position over the edge of the cake.

221

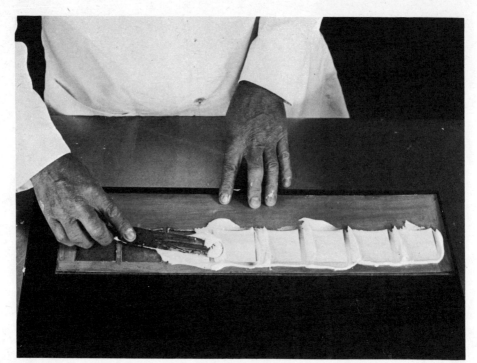

Fig. 218

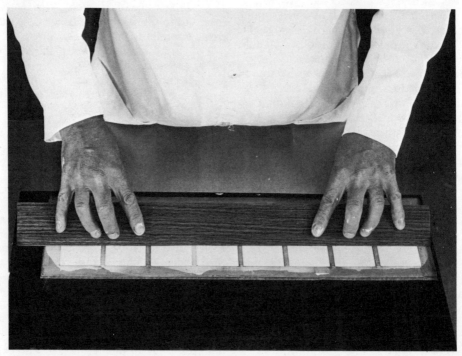

Fig. 219

222

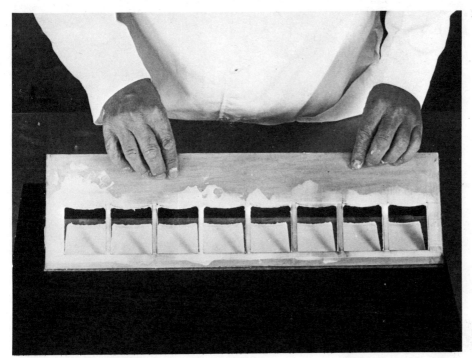

Fig. 220

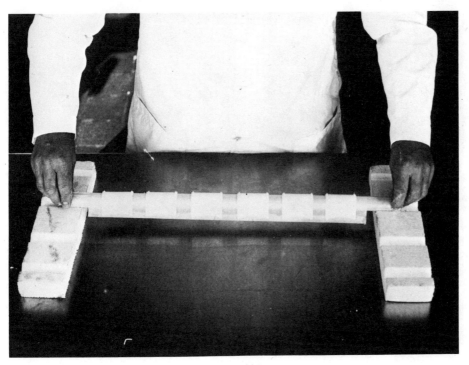

Fig. 221

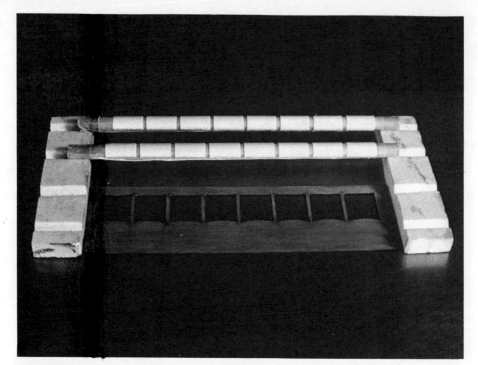

Fig. 222

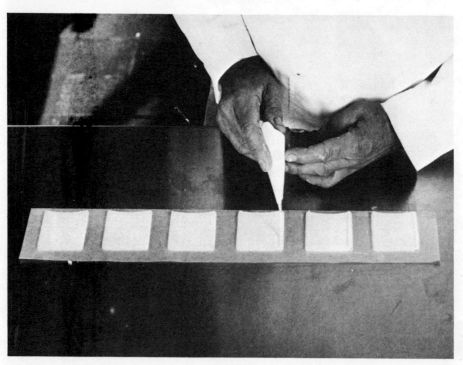

Fig. 223

224

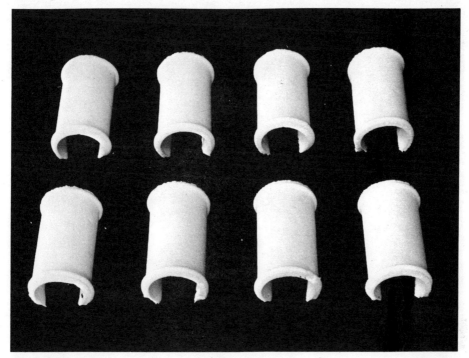

Fig. 224

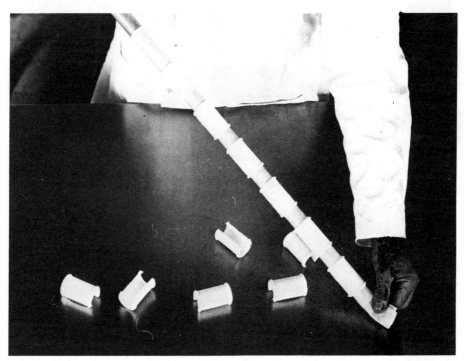

Fig. 225

225

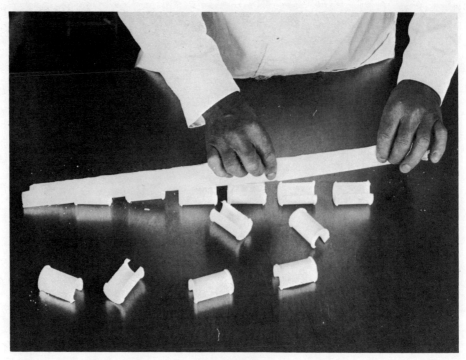

Fig. 226

Lilies

Using a similarly made stencil, lily petals are made (**Fig. 227**), the pieces being removed to set in a hollow template (**Fig. 228**). The shape of the lily is made by using a mould or narrow cup. A disc of waxed tissue is first placed into the bottom. In the centre of this a bulb of softened royal icing is piped and into this are placed the ends of the petals in six positions so that a lily shape is made. When set, the shape is mounted into the centre of the cake, either on a little silver coated plinth or a plain or fluted disc of sugar paste.

The cake shown in **Fig. 215** was very inexpensive to decorate using the minimum of inedible decorations and greatly reducing the time spent on decoration by the use of stencilled prefabricated pieces.

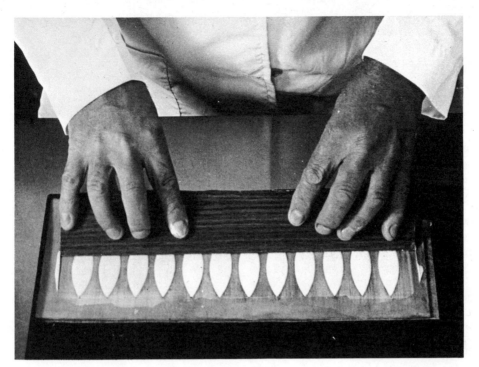

Fig. 227

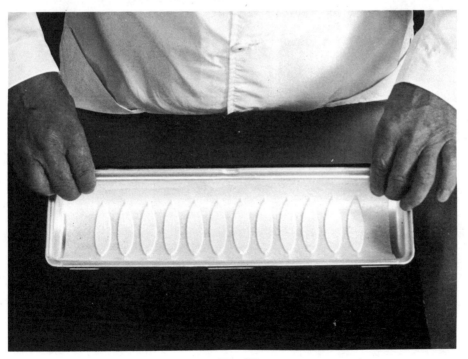

Fig. 228

Stencilled-off Pieces

Almost any shape can be stencilled and used for the decoration of borders of cakes and two more examples are given in **Figs. 229 and 230**.

Fig. 229 shows another circular cake, this time decorated with stencilled-off pieces. A decorative finish is made to these in the illustration by spinning over thinned royal icing either of the same or another colour. This has to be done immediately after stencilling and before the stencil mat is removed. This operation is shown in **Fig. 231**. It will not noticed that this mat contains four impressions, the stencilled pieces being available as an alternative to those used in **Fig. 230**. The cake in **Fig. 229** is simply finished by spinning the base with the same sugar as the stencilled-off pieces (**Fig. 232**) using a stencilled lily as a centrepiece and inedible decorations of silver leaves and ribbon.

Fig. 230 depicts a square cake similarly decorated. The off pieces have been left plain in this case, but outlined in a plain tube for a greater decorative effect and used for both top and bottom borders. The horseshoe was also stencilled, two sides being cemented with royal icing and the sides smoothed with a knife. Inedible leaves, shoes, horseshoes and silver band completes this simple design.

Many more examples of stencilled shapes suitable for the decoration of all types of decorative cakes could be given, but it is hoped that readers will use these ideas applied to wedding cakes to experiment on shapes which can be used for birthday, Christmas and all the other festive cakes which are traditionally decorated in royal icing.

Fig. 229

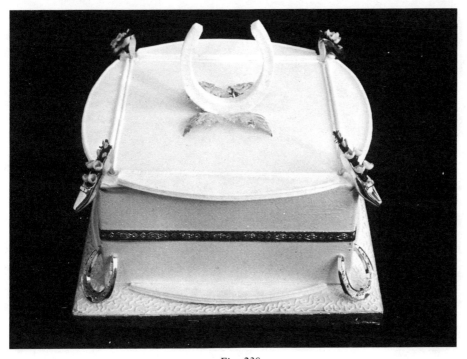

Fig. 230

229

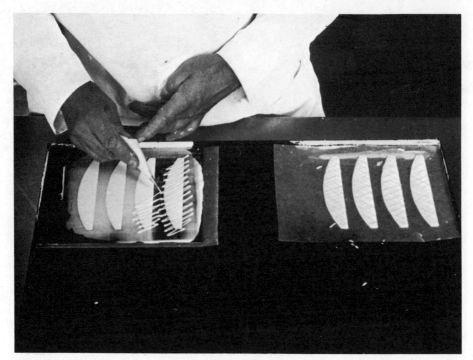

Fig. 231

Fig. 232

230

INDEX

Aerograph equipment, 3
  use of, 3, 21–29, 192
Almond paste, 6
  cut-outs, 18
  ring, 199
  tops,
    other, 190
    stencilled, 126

Basket Patterns, 169–171
Battenburgs, 134
Birthday Cakes, 203–209
Bulbs, 159
  piped, 155–161
Buttercream, 8
  slices, 138–145
Butterflies, stencilled, 29

Candles, 21, 54
Caramel Cutter, 3, 70
  use of, 19, 23, 47, 69, 140
Catkins, 52, 96
Celluloid stencil, 22
Cherry snowballs, 17, 100
Chicks, 27, 46
Chocolate, 7
  bakers, coating, 7
  block, 7
  buttons, 12
  couverture, 7
  curls, 14
  cut-outs, 14
  decor, 14
  flowers, 14
  holly leaves, 111
  icing, 8
  piping, 7
  run-outs, 14
  stars, 115
  unsweetened, 7
Christmas:
  cakes, 186–202
    sugar and almond-paste tops, 190
    use of stencil, 194
Christmas:
  chalet, 102
  gateaux, 100–117
  logs, 16
  torten, 120

trees, 21, 32, 106
Circular bands, 219–226
Colour, 1
  schemes, 202
  sprayed cakes, 192
Conserves, 9
Copper sheet, 22
Creams, various, 8
Crinoline lady, 172–174
Croquant, 8
Cross, the, 50
Crumbs, 10
Crystallized flowers, 10
  fruits, 9
Cut-out shapes, 21

Decorating materials, 5
Decorations, prefabricated, 11
Design, principles of, 1

Edibility, 2

Fancies, basic designs, 32–35
  basic shapes, 30
  chocolate and almond paste decor, 40
  Christmas, 44, 54
  Easter, 50
  flavour of, 40
  fruit, 42, 56
  stencilled tops, 46
Flowers, 20, 26
  crystallized, 9
  marbled, 178
  petals and leaves, 24
Fondant, uses of, 5
Fruit, fancies, 56
  gateaux, 98
  run-outs, 60
Fruits, sugar and preserved, 9

Ganache, 8
Gateaux, 62–117
  Christmas, 98, 100–117
  Easter, 66, 96
  fruit, 98
  Mother's day, 62–65

231

summer, 66
tops, stencilled, 84–99
  apple, 92
  apricot, 93
  banana, 93
  cherry, 90
  chocolate, 94
  Christmas, 87, 98–99
  Easter, 88, 96–97
  kirsch, 90
  lemon, 91
  Mother's day, 86
  orange, 91
  pear, 92
with stencilled inscriptions
  almond, 72
  cherry, 71
  chocolate, 78
  Christmas, 82
  coffee, 81
  date, 74
  date and walnut, 74
  Easter, 83
  hazelnut, 79
  kirsch, 71
  lemon, 75
  mandarin, 76
  nougat, 80
  orange, 75
  pineapple, 77
  praline, 80
  walnut, 73

Hand crimping, 134
Heather, 132
Holly leaves, 21, 54, 176

Icing aid, 2
  fudge, 8
  royal, 5
Inscriptions, 23

Jams, 9
Jellies, 9
Jelly decorations, 10

Lantern, 54
Layer cakes, 136–138
  stencilled tops, 146
Layout, 1
Leaves, 24
  marbled, 178
Lilies, 226–228

Marbled decoration
  gateaux, 148–154
  layer cakes, 138
Marbling leaves and flowers, 178
  line, 178
  other motifs, 182
Mistletoe, 54

Nougat, 8
Nut products, 8

Overpiping, 155

Parchment, 22
Piped decoration, 155
  crinoline lady, 172
Plaques, decorated, 184
Plastic film, 22
Praline, 8
Preserved fruits, 9

Rabbits, 50
Ribbon designs, 162
Robins, 28
Rope borders, 167, 208
  designs, 122
Royal icing, 5
Run-outs, Christmas cakes, 186
  decorated, 175–185
  rough, 184
  stippled, 184, 206
  sugar dusted, 184, 206
  two-colour, 180

Scrapers, 139, 205
Scrolls, 155, 164–166
Shells, 155–158
  designs, 157
Slices, 138–147
Snowballs, 17
  use of, 45, 101, 121
Stars, 54
Stencilled, almond paste tops
  fruit shapes, 132
  gateaux tops, 95
  run-outs, 60
  squares, 46
  torten, 126–133
Stencilling direct, 60
Stencils, 22
  two-colour, 48

Strawberries, 27
Sugar, 9
Sugar balls, 11
   border, 200
   buttons, 12
   paste cut-outs, 18

Tartan design, 133
Texture, 1
Thistle, design, 133
Torten, 118
   Christmas, 120, 150
   Easter, 124, 130
   markers, 5
   Mother's Day, 122
   section, 118
Turntable, precision, 2

Wedding cakes, 210–230
   bells, 211
   circular bands, 219
   cutting a wedge, 217
   decorations, 211
   heart shapes, 216
   lilies, 226–228
   pillars, 210
   proportions, 210
   rings, 213
   run-out borders, 213
   square, 215
   stencilled borders, 219
   stencilled-off pieces, 228–230
   traditional, 214

Yuletide logs, 109